Drought Management

Drought Management

Guru Sharan Das

RANDOM
RANDOM PUBLICATIONS
NEW DELHI (INDIA)

Drought Management

ISBN 978-93-5111-201-3

Published in 2014 in India by

RANDOM PUBLICATIONS

4376-A/4B, Gali Murari Lal, Ansari Road
New Delhi-110 002
Phone : +91-11-43580356, +91-11-23289044
e-mail: randomexports@gmail.com, sales@randompublications.com, info@randompublications.com

Type Setting by : Keystoneprintads, Delhi-110051
Printed at Thomson Press (India) Ltd

Preface

Drought is a natural hazard that differs from other hazards as it has a slow onset, evolves over months or even years and affects small pockets to a large regional expanse. Its onset and severity are often difficult to determine. As a result, there is a lack of urgency in response. Like other hazards, the impacts of drought span economic, environmental and social sectors and can be reduced through mitigation and preparedness.

Because droughts are a normal part of climate variability in virtually all regions, characterized by extended periods of water shortage, it is important to develop contextual plans to deal with them in a timely, systematic manner as they evolve.

The impacts of drought are diverse and often ripple through the economy. Thus, impacts are often referred to as either direct or indirect. A loss of yield resulting from drought is a director first-order impact of drought. However, the consequences of that impact (for example, loss of income, farm foreclosures, and government relief programs) are secondary or even tertiary impacts.

Risk based drought management is multifaceted and requires the involvement of a variety of stakeholders, and, from a drought management policy perspective, capacities in diverse ministries and national institutions are needed. Based on the identification of capacity needs from national to local levels, the initiative aims to develop such capacities to enable countries develop pro-active, risk-based drought management policies and overcome prevailing structures of reactive and post-hazard management strategies.

Drought risk management (DRM) is the concept and practice to avoid, lessen or transfer the adverse effects of drought hazards and the potential impacts of disaster through activities and measures for prevention, mitigation and preparedness.

It is a systematic process of using administrative directives, organizations and operational skills and capacities implement strategies, policies and improving coping capacities.

In addition the inclusion of long term measures like water conservation projects such as aquifer recharge and the constructs of reservoirs, has also been taken care of in this book.

I thank all members of my team who have helped in the preparation of the book. My special thanks go to "Random Publications" who have published the book.

*– **Guru Sharan Das***

Contents

Preface .... *v-vi*

1. Introduction **1**

Drought as a Natural Hazard 1
Drought Management in India 6

2. Drought Risk Reduction and Management **8**

Drought Risk Management 8
Drought Risk Management Projects at a Glance 9
Drought Risk Management Contracts 9
Integrating Disaster Risk Reduction in Development 19
Disaster Management at Regional and Sub-Regional Levels 20
Uncertainty, Risk and Risk Mitigation 27

3. Drought Mitigation Interventions in the Pastoral Livestock Sector **35**

Identify the Critical Sectors in each State that are Drought Vulnerable 43
Sustainable Livelihoods for Poverty and Disaster Risk Reduction 46

4. Drought Preventability and Prediction in a Changing Climate **72**

Introduction and Motivation 72
User Requirements for Drought Prediction Information 73
Effective Communication of Drought Information 79
Drought Environments 99

5. Drought Hazard Assessment and Mapping **106**

Introduction 106
Assessment 109
Land Use and Land Management Practices 118

Infrastructure 126
Impact and Vulnerability of Drought 130
Drought Impacts 131
Analysis of Areas at Risk 140
Drought Mapping 147
Identification of Droughts 150

6. Drought Management for Pastoral Livelihoods and Policy 154

Introduction and Background 154
Drought Management and Relief 155
Public Policy and Drought Management in Agropastoral Systems 163
Drought Policy Responses and Policy Debates 176

7. The Impacts of Water Management During Drought 183

Improving River Management During Drought 185
Consecutive Droughts of Rajasthan Through Participatory Water Resource Development 188
Impact of Climate Change on Water Resources 194

8. Drought Resistance 204

Introduction 204
Characterization of the Target Environment and Drought Definition 207
Drought Resistance in the Home Landscape 209
The Drought Breeding Strategies 215
Integrated Drought Management Options 218

9. Remote Sensing for Drought Monitoring 219

Introduction 219
Meteorological Drought Analysis Using GIS 221
Impact of Drought Using Remote Sensing 223

10. Drought and Sustainable Agriculture 248

Background 248
Demographic and Socioeconomic Characteristic of Fadama farmers 252
Sustaining Agriculture in the Era of Climate Change in India 267
Sustainable Development of Indian Agriculture 281

***Index* 290**

1

Introduction

Drought is a subtle, insidious natural hazard that is a normal part of the climate of virtually all regions of the world. Its occurrence results in a myriad of economic, social, and environmental impacts in developed as well as developing nations, although the characteristics of its impacts differ considerably between the two settings. Drought is considered by many to be the most complex but least understood of all natural hazards, affecting more people than any other. It is a normal feature of climate and its recurrence is inevitable. However, there remains much confusion about its characteristics. It is precisely this confusion that explains, to some extent, the lack of emphasis on proactive drought management efforts in most parts of the world. Through an improved understanding of the inevitability and characteristics of drought, as well as its differences from other natural hazards, scientists, policymakers, and the public will be better equipped to establish much-needed policies and plans whereby future vulnerability to drought can be reduced.

DROUGHT AS A NATURAL HAZARD

Drought differs from other natural hazards in several ways. First, drought is a slow-onset, creeping natural hazard. Its effects often accumulate slowly over a considerable period of time and may linger for years after the termination of the drought event. Second, the absence of a precise and universally accepted definition of drought adds to the confusion about whether a drought exists and, if it does, its degree of severity. Third, the impacts of,

During the Dust Bowl of the 1930s, nearly two-thirds of the U.S. land area experienced severe to extreme drought conditions. This ruined farm is buried under layers of dirt and sand blown across the landscape. Drought are nonstructural and typically are spread over a larger geographical area than are damages resulting from other natural hazards. These characteristics of drought have hindered the development of accurate, reliable, and timely estimates of severity and impacts and, ultimately, the formulation of effective drought preparedness plans.

DROUGHT CHARACTERISTICS AND DEFINITION

Drought results from a deficiency of precipitation from statistically normal (long-term average) amounts that, when extended over a season or especially over a longer period of time, is insufficient to meet the demands of human activities. All types of drought originate from a deficiency of precipitation that results in water shortages for some activity (such as crop production) or for some group (such as farmers).

Droughts differ from one another in three essential characteristics: intensity, duration, and spatial coverage. Moreover, many disciplinary perspectives of drought exist. Because of these numerous and diverse disciplinary views, considerable confusion often exists over exactly what constitutes a drought. Regardless of such disparate views, the overriding feature of drought is its negative impacts on people and the environment.

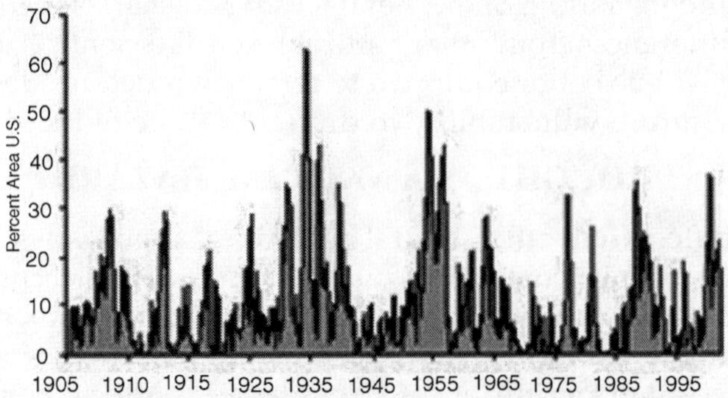

This graph shows the percent area of the United States in severe and extreme drought during the twentieth century.

TYPES OF DROUGHT

Drought is normally grouped by type: meteorological, hydrological, agricultural, and socioeconomic. The impacts associated with drought usually take 3 months or more to develop, but this time period can vary considerably, depending on the timing of the initiation of the precipitation deficiency.

Meteorological

Meteorological drought is expressed solely on the basis of the degree of dryness in comparison to some normal or average amount and the duration of the dry period. Thus, intensity and duration are the key characteristics of this type of drought.

Agricultural

Agriculture is usually the first economic sector to be affected by drought because soil moisture content is often quickly depleted, especially if the period of moisture deficiency is associated with high temperatures and windy conditions. Agricultural drought links various characteristics of meteorological drought to agricultural impacts, focusing on precipitation shortages, differences between actual and potential evapotranspiration, and soil water deficits. A thorough definition of agricultural drought should account for the variable susceptibility of crops at different stages of development.

Hydrological

Hydrological droughts are associated with the effects of periods of precipitation shortfall on surface or subsurface water supply (for example, streamflow, reservoir and lake levels, and groundwater) rather than with precipitation shortfalls.

Hydrological droughts usually lag the occurrence of meteorological and agricultural droughts because more time elapses before precipitation deficiencies are detected in reservoirs, groundwater, and other components of the hydrologic system. As a result, impacts of hydrological drought are out of phase with impacts of other drought types. Also, water in hydrological storage systems such as reservoirs, rivers, and groundwater often is used for multiple and competing purposes, further complicating the sequence and quantification of impacts. Water uses affected by drought can include purposes as varied as power generation, flood control, irrigation, drinking water, industry, and recreation.

Socioeconomic

Socioeconomic drought associates the supply and demand of some economic good or service with elements of meteorological, hydrological, and agricultural drought.

In socioeconomic drought, deficiencies of precipitation are linked directly to the supply of some commodity or economic good (for example, water, hay, or hydroelectric power). Increases in population can alter substantially the demand for these economic goods over time. The incidence of socioeconomic drought can increase because of a change in the frequency of meteorological drought, a change in societal vulnerability to water shortages, or both. For example, poor land-use practices such as overgrazing can decrease animal carrying capacity and increase soil erosion, which exacerbates the impacts of, and vulnerability to, future droughts.

THE IMPACTS OF DROUGHT

The impacts of drought are diverse and often ripple through the economy. Thus, impacts are often referred to as either direct or indirect. A loss of yield resulting from drought is a direct or first-order impact of drought. However, the consequences of that impact (for example, loss of income, farm foreclosures, and government relief programs) are secondary or even tertiary impacts.

The impacts of drought appear to be increasing in both developing and developed countries, which in many cases reflects the persistence of nonsustainable development and population growth. Lessening the impacts of future drought events will require nations to pursue development of drought policies that emphasize a wide range of risk management techniques, including improved monitoring and early warning systems, preparedness plans, and appropriate mitigation actions and programs. The impacts of drought that must be addressed can be classified into one of three principal types: economic, environmental, and social.

Economic Losses

Economic impacts range from direct losses in the broad agricultural and agriculturally related sectors (including forestry and fishing), to losses in recreation, transportation, banking, and energy sectors. Other economic impacts would include added unemployment and loss of revenue to local, state, and federal government.

Environmental Impacts

Environmental losses include damages to plant and animal species, wildlife habitat, and air and water quality; forest and range fires; degradation of landscape quality; and soil erosion. These losses are difficult to quantify, but growing public awareness and concern for environmental quality has forced public officials to focus greater attention on them.

Impacts on Society

Social impacts mainly involve public safety, health, conflicts between water users, and inequities in the distribution of impacts and disaster relief programs. As with all natural hazards, the economic impacts of drought are highly variable within and between economic sectors and geographic regions, producing a complex assortment of winners and losers with the occurrence of each disaster.

DROUGHT PREPAREDNESS AND MITIGATION

Drought is considered by many people to be strictly a natural or physical event. This view of drought provides little, if any, opportunity to alter the impact of drought through the application of appropriate drought management techniques. In reality, drought has both physical and social components, and it is essential that water managers and decisionmakers understand

A prolonged drought in the eastern United States in 1999 lowered the water level of the Lehigh River approximately 5 feet. The much-reduced flow affected municipal and industrial uses along the river, including the Bethlehem Steel Plant (background). Both components if they are to make progress in reducing the risks associated with drought for future generations.

It is critical that the people inhabiting each geographic region understand their exposure to the drought hazard: for example, the probability of drought occurrence at various severity levels. However, the risks associated with drought for any region are products of both the region's exposure to the event and the vulnerability of its society to a drought at that point in time. Vulnerability, unlike the natural event, is determined by varied social factors. Examples include:

- Population changes;
- Population shifts (region to region and rural to urban);
- Demographic characteristics;
- Environmental awareness (or lack thereof);
- Level of technology;
- Wisdom and applicability of government policies;
- Land management practices; and
- Social behavior.

These factors change over time and thus vulnerability is likely to increase or decrease in response to these changes. Subsequent droughts in the same region will have different effects, even if they are identical in intensity, duration, and spatial characteristics, because societal characteristics will have changed. However, much can be done to lessen societal vulnerability to drought through the development of preparedness plans that emphasize risk management and the adoption of appropriate mitigation actions and programs.

Management Alternatives During Drought

Many of the management alternatives available to water suppliers to prepare for and respond to drought events fall into two broad categories: demand management and supply augmentation.

Demand Management:

- Public information and education campaigns
- Emergency conservation programs
- Water service restrictions
- Restrictions on nonessential uses of water
- Prohibition of selected commercial uses
- Drought emergency pricing
- Water rationing programs
- Augmentation
- Improvements in water systems (for example, leak detection, lining of transmission canals)
- Emergency sources of supply (for example, emergency interconnections, drilling new wells)
- Management of available water resources (for example, emergency water banks, overdrafting of groundwater aquifers)
- Search for new supplies of water

In summary, increased emphasis can and should be placed on drought mitigation and preparedness, as well as prediction and early warning capabilities, if society is to reduce the economic and environmental damages associated with drought and its personal hardships. This will require interdisciplinary cooperation and a collaborative effort with policymakers at all levels.

DROUGHT MANAGEMENT IN INDIA

The drought-prone area assessed in the country is of the order of 51.12 Mha. The planning and management of the effects of drought appear to have a low priority due to associated randomness and uncertainty in defining the start and end of droughts. Further, most of the drought planning and management schemes are generally launched after persisting drought conditions.

The traditional system of drought monitoring and estimating losses by crop cutting needs replacement with real time remote sensing, GIS, GPS and modeling techniques for ensuring transparency and quick response. Scope of losses may be extended to groundwater depletion, damage to perennial trees, plantations, orchards and depletion infertility of livestock. Food, Toddler, agricultural inputs and water banks may be established in vulnerable areas instead of their storage in surplus regions to avoid transport bottlenecks during the drought. Robust and rainfall independent off-farm livelihood opportunities may be targeted in the drought mitigation strategy. Conjunctive use of surface and groundwater, aquifer recharge and watershed management with community participation is another important policy paradigm shift to be internalized fully. After normal rainfall resumes there is a rapid decrease of government and public interest in drought-planning schemes. Most of the time the execution of the drought management scheme is based on the administrative units, while planning of water resources is based on basin scale.

Therefore, an integrated basin development approach is required to be developed and implemented for preparing the drought management plan before, during and after the occurrence of the drought. In this regard, there is a need for the development of the decision support systems (DSS) for the monitoring and management of the drought on basin scale utilizing the advanced capabilities of remote sensing, geographical information system and knowledge based systems.

The DSS should also provide support to the decision makers for providing the information at different spatial and temporal scales. It would help them in taking the required management measures in the drought- prone areas for different administrative units. In the drought-prone areas, publication campaign may be launched for water conservation with the help of electronic and print media. Necessary steps may be taken at political, administrative and technical levels to encourage people participation in the drought management for optimum utilization of the available water supply to meet the demands. Strengthening of R&D and capacity building in terms of emerging information technologies and issues of damages is also called upon to bring in resilience in the drought coping strategies.

2

Drought Risk Reduction and Management

DROUGHT RISK MANAGEMENT

People who live in dry areas are vulnerable to disasters of various kinds. They are subject to recurrent droughts, and when the rains come, they are often affected by serious floods. In the past, crisis preparedness and management often focused on man-made disasters and acute natural disasters. Recurrent exposure to natural hazards, especially drought, has been largely ignored. This is changing, and UNDP-DDC is helping the change to come about. Drought risk management (DRM) is the concept and practice to avoid, lessen or transfer the adverse effects of drought hazards and the potential impacts of disaster through activities and measures for prevention, mitigation and preparedness. It is a systematic process of using administrative directives, organizations and operational skills and capacities to implement strategies, policies and improving coping capacities.

In recognition of the multiplicity of drought challenges in the context of uncertainties surrounding climate change, UNDP-DDC is focusing on building long-term resilience to climate shocks and change as well as mitigating immediate disaster risks and impacts. This is being undertaken through mainstreaming DRM into development planning and practices with the financial support from the European Union, the Government of Finland and the Government of Japan. The country projects assist national as well as local counterparts to shift from reactive to proactive drought management approaches by identifying and implementing options for reducing drought vulnerability in an integrated manner.

DROUGHT RISK MANAGEMENT PROJECTS AT A GLANCE

In Mozambique, the project supported the organization of exchange visits/ study tours and training workshop, where the national bushfire prevention plan was presented by local authorities and its implementation discussed with community leaders. Experiences and lesson learnt on bush fire prevention and control from Nampula Province, which is considered as the best practice in Mozambique, was also shared.

Furthermore a Training of Trainers was carried out, where personnel from the various District Administrations responsible for building capacity of communities in arid and semi-arid areas have been trained in promotion of efficient use of natural resources. Through this process communities have now been empowered to improve productivity of drought resistant products/ goods. Farmers have also been trained in sustainable apiculture practices and provided needed equipment (modern beehives) to continue developing this alternative livelihoods option.

The project in Namibia, which is still in its preparatory phase, intends to develop materials and tools to be used for building capacity of communities to prevent and mitigate drought risk in the drylands. The initiative will collaborate with the other ongoing programmes and initiatives operating in the fields of climate change adaptation and disaster risk management. Tools and materials developed will be applied in two communities to fill in DRM capacity gaps identified in the baseline assessments. The objective is to raise awareness on disaster risk management as well as to enhance the resilience of the target communities to address future drought and other climate risks. Lesson learnt and success stories from this process will be identified, documented and shared.

In Syria, the project seeks to enhance community adaptation to drought within the framework of the project "Integrated Community Development for Scaling-up the MDGs". This project aims at improving MDG related service delivery through an integrated community development approach; it targets 6 villages in the North-Eastern Region which was hit by severe drought. Programme support is used to build community resilience to drought through action at two levels: 1) the promotion of appropriate technologies/practices in land and water management; and 2) the development of vocational and business skills and the establishment of small and medium enterprises.

DROUGHT RISK MANAGEMENT CONTRACTS

Recent developments in international risk markets imply that the Government can buy an index-based weather risk management contract to transfer the financial risk of severe and catastrophic national drought that adversely impacts the Government's budget to the international risk markets

(Hess and Syroka, 2005; Ibarra, et al., 2005; World Bank, 2005; Syroka and Wilcox, 2006; Government of Malawi and World Bank, 2007a; Government of Malawi and World Bank, 2007b; WFP, 2007; also see Box 1). Index-based risk transfer approaches are not insurance in the traditional sense where the insured party is compensated for the precise monetary loss that can be proved to have occurred. Rather the protection is based on the performance of a specified index during the contract period in relation to a specified trigger, where the index is designed to correlate as closely as possible with the underlying risk. Payouts are automatically made whenever this index crosses the specified contract threshold at the end of the contract period, without the need of loss assessments and checks to confirm the occurrence and magnitude of an actual loss. Often these contracts are legally structured as derivative contracts (rather than insurance contracts) although they still perform an insurance function.

The counterparties to such transactions, the entities to whom the risk is transferred for a premium charge, are actors in the weather risk market. The weather market has grown exponentially in recent years, with over US$100 billion of risk transferred since its inception in 1997 (WRMA 2007). Weather market players, from both the reinsurance and financial communities are extremely interested in new, uncorrelated risks, such as Malawi's, that can diversify and therefore enhance commercial risk portfolios.

In order to implement a successful weather risk transfer program, however, the primary technical precondition is that the data used to construct the underlying weather indexes must adhere to strict quality requirements, including: reliable and trustworthy on-going daily collection and reporting procedures; daily quality control; an independent source of data for verification; a long, clean, and internally consistent historical record to allow for a proper actuarial analysis of the weather risks involved—at least 25-30 years of daily data are ideally required, with less than 3-5 percent missing. The strict nature of these criteria is to ensure the underlying weather data used is objective, easily accessible and independently verifiable, thus ensuring that the balance of information between the two entities is shared equally and the nature of the risk being transferred is transparently communicated.

Experience shows that in most countries, including least developed countries (LDCs), there is often enough weather stations and data to consider an index-based risk management approach at the national level with a little additional investment. The Malawi Meteorological Services Department's rainfall data is of excellent quality, therefore satisfying a key prerequisite for risk transfer.

An index-based drought risk management contract offers several advantages for Malawi. One advantage of weather-indexed products, over products based on more subjective loss assessments, is that they create market access opportunities: the risk can be transferred from the country in question to the commercial portfolios of actors that actively seek this risk. Secondly, by using a proxy measure of risk, such as rainfall that is available on a real-time basis, a contract would secure timely and guaranteed funds for the Government if a contractually specified severe and catastrophic shortfall in precipitation occurs during the agricultural season. These products can be settled as soon as the data is received and verified, without having to rely on time consuming field-level assessments and often subjective loss adjustment processes to assess losses, such as those required for traditional insurance products, or in the case of traditional appeal-based humanitarian aid, on lengthy needs assessments.

Such a predictable source of early emergency financing in a time of crisis – together with improvements in the country's early warning system – will give the Government more flexibility in its drought preparedness and contingency planning, allowing an earlier response to shocks and saving overall costs: (a) directly, through more timely, informed and optimal operational decision making and; (b) indirectly, though a swift and sound response that reduces the overall negative developmental impact of a drought event and protects against the financial disruption of other Government programs. By securing supplemental resources, through risk transfer to the international risk markets, the Government requires less contingency capital within its budget, which can be more effectively employed on other development programs. In addition, minimizing Government dependence on traditional appeals-based emergency financing which is unpredictable and often untimely, promises greater dignity for the countries vulnerable to weather shocks. Finally, discovering the market price of Malawian drought risk will assist the Government in making superior investment decisions with respect to managing and mitigating this risk within its portfolio and smoothing out drought-related expenditures across good and bad years.

It is important to highlight that index-based weather risk management instruments should be viewed as just one of a growing number of tools and options in the Government's arsenal for managing food security and creating fiscal stability in the face of exogenous shocks. In order to maximize the cost benefits of timely and predictable financing, and reveal the value of transferring only severe and catastrophic drought risk, the program needs to be implemented with a comprehensive maize risk management strategy and contingency plan.

In terms of disadvantages, one of the biggest limitations of index-based risk management products is the "basis risk", defined as the potential mismatch between the index-based contract payout and the actual maize production losses in country, such that the payout does not adequately indemnify the government for their losses. Basis risk can occur both ways: Malawi could experience a maize production shortfall that is not picked up by the index and therefore does not trigger a payout; or Malawi could experience an overall national maize production surplus but the contract still triggers a payout.

Furthermore, in the context of Malawi for example, not all maize production deficits are caused by drought: flooding, civil strife, poor farm-management, pest infestations and inadequate seed and fertilizer supplies may be as important as deficit rainfall in triggering food emergency situations. These risks are not captured by an index based on rainfall and cannot be objectively indexed for risk transfer. This limitation must be clearly communicated to all parties, and the underlying indices should be jointly developed with government partners to ensure that the index correlates as well as possible with the underlying risk, and to ensure appropriate country ownership and understanding of the approach. For the same reasons, wherever possible, the index should also be adapted from locally developed models that have been used by countries for some time. Nevertheless, basis risk, whether real or perceived, means that index-based product offerings must always be accompanied by other risk management or mitigation opportunities, and contingency plans, if the full value of these products, and of risk transfer, is to be realized.

A further disadvantage to such an approach is that the contract has an upfront, annual non-refundable cost, which is additional to the cost of a full-scale Government drought response. Donors often provide additional financial resources in response to a humanitarian crisis; hence, there may be a financial loss for the Government if it takes on the cost of the insurance premium annually. Such a perverse incentive against an ex-ante approach to risk management should be discussed with development partners to seek support in the payment of at least part of the premium.

DEVELOPING AN INDEX-BASED WEATHER DERIVATIVE CONTRACT TO MANAGE DROUGHT RISK

The very high correlation between maize yields and rainfall, and the substantial role played by maize production in the Malawian economy, suggest that a weather risk management contract for Malawi could be designed based on a maize production index.

The Malawi Maize Index

The success of a weather risk management program for the Government of Malawi depends on the design of the underlying rainfall index on which the contract is based. As outlined above, an index-based strategy does not cover the actual production loss of maize; instead it covers the shortfall in rainfall needed for adequate maize growth and development during a drought year. Because rainfall and maize yields are highly correlated, changes in rainfall – its cumulative amount and distribution – can act as a proxy for maize losses. Risk mitigating payouts are made from the market counterparty if the index crosses a specified trigger level at the end of the contract period, indicating situations where a loss is most likely to have occurred, i.e., payouts are made based on the index and not on the actual loss itself.

The index proposed for Malawi – the Malawi Maize Index (MMI) – is constructed using rainfall data from 23 weather stations throughout the country. The MMI is based on the Malawi Meteorological Services Department's (MMSD) national maize yield assessment model, used by the Government since 1992 to produce national maize production forecasts each February. The MMSD's national maize yield assessment model is based on 75 weather stations and rain gauges throughout the country. Because only 23 stations within the Malawi network satisfy the strict criteria risk for transfer to the weather risk market, the MMI only uses these 23 primary stations. Other than this difference, the methodologies and input parameters for the two models are precisely the same.

The MMSD (and MMI) model is a modified version of the FAO's Water Requirement Satisfaction Index (WRSI) adapted to Malawian conditions, and uses daily rainfall as the only varying input to predict maize yields and therefore production throughout the country. In this way the model, and therefore the index, isolates the impact of only rainfall variability on maize production. The rainfall and agricultural season in Malawi starts in October and finishes in April the following year, therefore daily rainfall from October to April is used to construct the MMI. Based on a water balance calculation, the model captures not only the total amount of rainfall received at each station, but also its distribution during the agricultural season and how these rainfall deficits impact maize yields. By using such a model, a contract can be structured to reflect conditions which would impact national maize production and therefore food security.

The inter-annual variations in the MMI have a correlation coefficient of 80 percent with inter-annual variations in historical national maize yields for 1984-2007 as early as three months before harvest and 93 percent with inter-annual variations the MMSD's national maize production forecasting model for 2000-2007.

The limitations of the MMI model, as with all index-based strategies, relate to basis risk, as discussed above. Thus, payouts are based on rainfall and not on actual maize production levels in the country. With only 23 weather stations, evaluating rainfall levels for the country as a whole is still rather crude. One part of the country may experience drought while another experiences favorable rainfall. Similarly, farmers residing near a weather station may experience very different levels of rainfall and crop growth than farmers further away from the weather station. As such, although the 23 weather stations are distributed widely across the country, the current MMI provides a gross measure of drought for the country as a whole and is not appropriate for regional, localized or mild drought events. As more stations with automatic communication links become available, the index will improve further and can become more flexible. Investments to improve the MMSD's model to better capture the impact of erratic and excess rainfall on maize yields will also improve the MMI and the accuracy of payouts when rainfall-related losses occur.

To reinforce the point that the index is not a perfect indicator of national maize production, which is also impacted by other factors that are not captured by the index, the MMI is expressed in terms of the percentage deviation of the index from the long-term average, rather than absolute metric tones of maize produced. Had it been in place, a derivative contract based on such an index would have triggered cash payouts to the Government most recently in 2005.

Going back to 1960, such a contract would have also made payouts in 1995, 1994, 1992, 1980 and 1970. In all such cases, these payouts could have been used, for example, to finance maize purchases or price risk management instruments to cap the price of maize imports, or other humanitarian interventions.

The Contract Structure

A simple example will clarify the operation of such a contract. Consider that the Government enters into a weather derivative contract structure based on the MMI. The contract has a trigger level of 95 percent, i.e., a payout is only made if at the end of the agricultural season in April the MMI is calculated to be below the 95 percent trigger level (meaning that the season's MMI value is less than 95 percent of its long term average, implying the total national maize production is also down due to deficit rainfall-related losses). If the MMI is above 95 percent at the end of April no payment is made.

The contract is designed such that if the MMI is below the 95 percent threshold the Government will receive US$500,000 for every percent point that the MMI is below 95 percent, up to a maximum payout limit of US$20 million. This approximates to an additional US$10 per each MT that the national maize production drops below average.

Such a contract would have paid out US$16.2 million in 1992, US$16.0 million in 1994, US$10.1 million in 1995 and most recently US$1.8 million in 2005.

The average payout, or expected loss of the contract, over 45 years is US$1.4 million. For entering into this contract and transferring this expected loss to the market and securing the right to access contingency funds of up to US$20 million if there is a drought-related crisis as measured by the MMI, the Government must pay an annual premium of approximately US$2.5 million. From a donor perspective this is can be viewed as an opportunity to leverage US$2.5 million premium contribution into a US$20 million drought-response contribution in a worst case scenario where a severe and catastrophic shortfall in precipitation has occurred. More importantly, as discussed above, such an ex-ante commitment enables the Government to build a drought contingency plan well in advance of a drought event around the availability of the contract payout, which will ensure a better planned, more efficient and cost-effective disaster response.

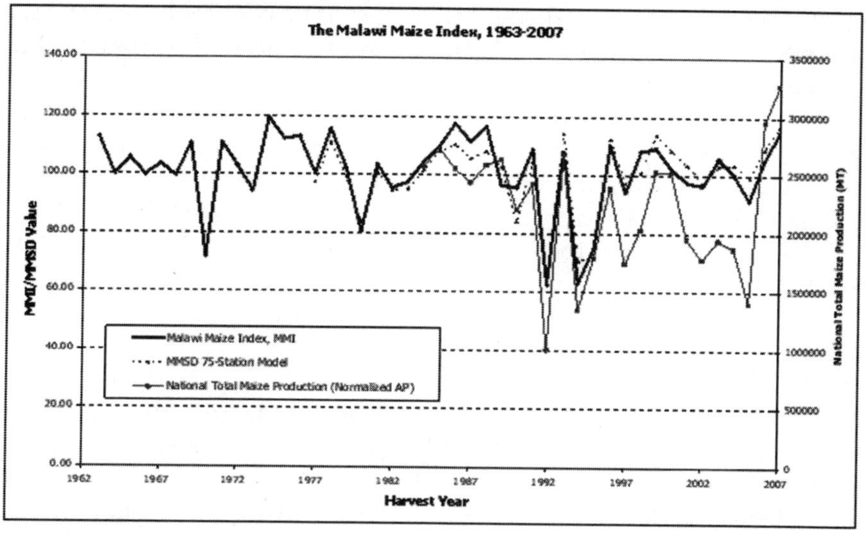

Fig. The MMI and MMSD model-predicted production deviations from average versus historical total Malawi national maize production (MT)

The expected (annual average) payout for such a contract would vary with the trigger level. When establishing a price for a weather risk management instrument, market providers will take into consideration their own risk appetite, business imperatives, cost of risk and operational costs. While there are a variety of methodologies for pricing, in general the pricing for all contracts will contain an element of expected loss, plus some loading or risk margin that corresponds to a capital reserve charge required to underwrite the risk at a target level for the business, as well as administrative costs.

Table. Expected premium and Payouts at Different trigger Levels (for US$20 Million Payout Limit)

Trigger Level	Expected Loss	Indicative Premium	Rate-on-Line
95	US$ 1,400,000	US$ 2,500,000	13%
90	US$ 1,100,000	US$ 2,000,000	10%
85	US$ 800,000	US$ 1,600,000	8%
80	US$ 500,000	US$ 1,100,000	6%

Note: This table provides tentative estimates of premium rates; only through a competitive market tender at various trigger levels will the true price of Malawian weather risk be discovered.

Therefore in general the premium charge for a contract can be broken down as follows:

Premium = Expected Loss + Risk Margin + Administrative Costs

Expected loss is the average payout of the contract in any given season. In some years payouts in excess of the expected loss can occur and the risk-taker must be compensated, through the risk margin, for this uncertainty and the internal provisions that must be made in order to honor these potentially large payouts. The values of the expected loss and the risk margin must be established from historical weather data. The approach for determining the risk loading over the expected loss differs from risk-taker to risk-taker and many use a combination of methods to deter¬mine the risk margin included. A sensible pricing methodology uses a risk measure such as the Value-at-Risk (VaR) of the contract to determine the risk margin. A VaR calculation is aimed at determining the loss that will not be exceeded at some specified level of confidence, often set at 99 percent. It targets the potential extreme negative deviations in payouts for the risk taker – regulators and rating agencies, for example, use such tail-risk measures in determining the capital a bank, (re)insurer or corporation is required to hold to reflect the risks it is bearing – and therefore the associated capital charge for taking on the risk. Administrative costs are essentially the costs for the provider to run the business including charges for data, office costs, taxes and brokerage charges if necessary.

The indicative premiums in Table do not include administrative expenses and provide a simple illustrative example, using the VaR pricing approach, of how premiums may expect to vary with MMI thresholds using conservative risk loadings observed in the international weather market. It is important to note that the premium is always greater than the expected loss; that is, Malawi would always pay more for the insurance than it expects to get back in payouts over the longer term. However the additional value for Malawi of this contract that is not captured in these calculations is the financial savings the Government would make by having access to guaranteed funds as early as May, enabling a timely response to a developing crisis. If these leveraging effects of a weather derivative payout are taken into account, over the long-term the cost benefit analysis of an annual market premium charge are likely to be significantly more favourable for Malawi.

For example, following the drought and low harvest in May 2005, Malawi imported between 200,000-300,000 MT of maize into the country for distribution during the lean months that followed in January-March 2006. Much of this maize was purchased on the spot market in South Africa in October-November 2008 and transported into the country for immediate distribution. Had Malawi been in the position to purchase maize in June-July 2005, or ideally to purchase price risk management instruments that capped the maize import price for the months ahead, it would have saved approximately US$80-110 per MT imported, using conservative estimates. Calculations show this saving means a weather derivative payout directed into such investment in June would have had contributed up to a five-fold saving for the Government. Assuming similar market dynamics hold in other deficit production years, the saving gains through using a weather derivative payout would exceed the additional risk margin charged for the contract in the long-term.

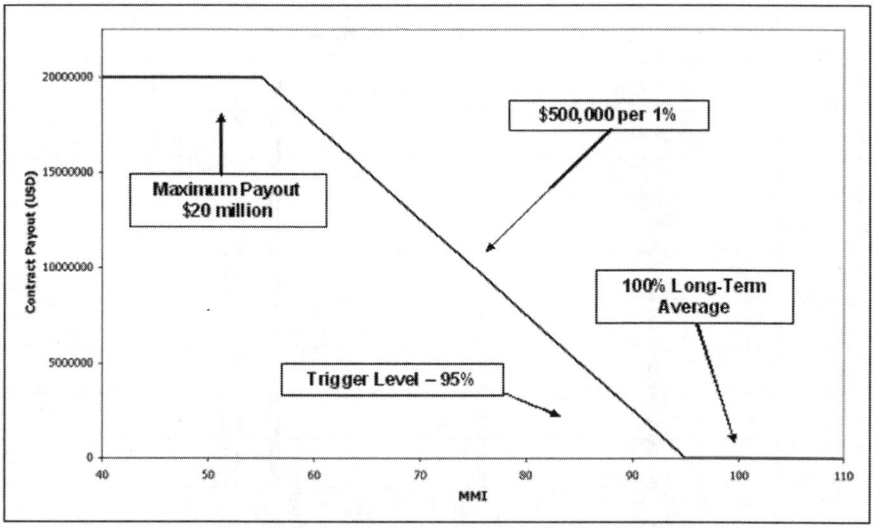

Fig. An example of a hypothetical MMI-based weather derivative payout structure, 95 percent trigger level

The development of this financial instrument could ultimately enhance the Government's response to drought-induced food shortages, and mitigate the impact of weather-related shocks on macroeconomic performance, thereby facilitating performance, thereby facilitating investments, economic growth, and poverty reduction. The next step is for the Government of Malawi to enter into an initial pilot transaction, which may take place in 2008/09. In fact, several piloting seasons will be necessary to understand the scope and limitations of weather derivative contract, and its role within the Government's strategy, contingency planning and operational drought response.

DFID, the EU and USAID have signaled their interest in supporting such a pilot program and providing funding to pay for the risk-transfer premium. The cost of the 2004/2005 drought response in Malawi was over US$200 million for the Government and the pilot should not aim to cover the entire financial magnitude of such a potential drought response in the country. Rather it should provide coverage for a defined percentage of this risk as part of a broader maize risk management program and will test the feasibility and cost of transferring Malawi's drought risk to the international risk markets. The objective of piloting is not only to secure limited drought financing in a timely and objective manner but moreover to establish a price for Malawian drought risk. The actual market price of Malawi drought risk will not be known until an official competitive market tender for the risk is launched as part of a real transaction. In the future as more and more transactions of this type occur, and as commercial portfolios become more diversified and the market familiar with such new risk and clients, it is expected the average cost of risk transfer will also come down.

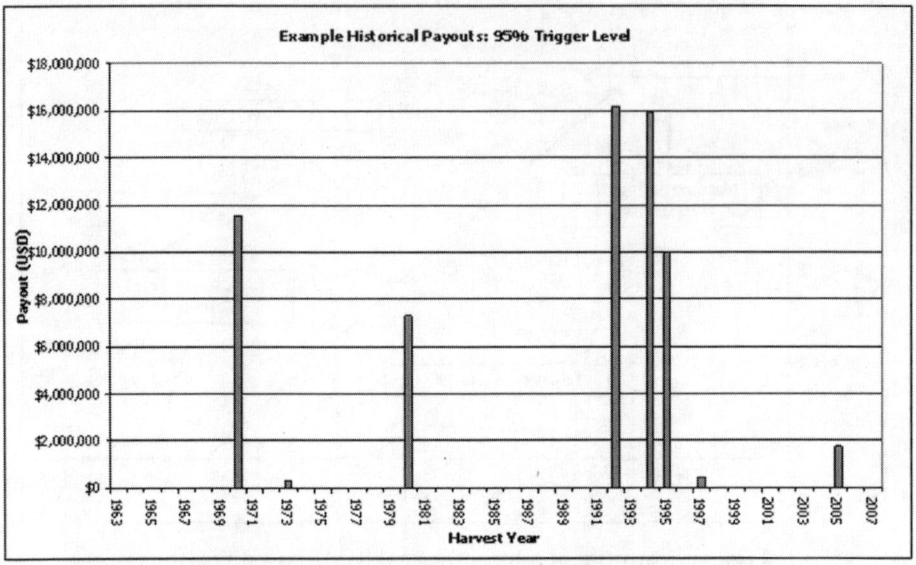

Fig. An example of historical payouts at the 95 percent trigger level

Climatic shocks are frequent in Malawi and constitute a major impediment to economic growth and poverty reduction. Moreover, global climate change models suggest that future volatility in weather patterns in Sub-Saharan Africa are likely to become even more pronounced. Their impact goes far beyond the initial need for emergency relief, and the resulting macroeconomic instability has presented a majo7r constraint to growth and poverty reduction in Malawi. This paper describes an innovative financial instrument which could allow the Government of Malawi to transfer the financial risk of severe and catastrophic national drought to the international risk markets.

The development of an index-based drought derivative contract will create a new opportunity for Government to financially manage its budgetary exposure to maize production risk using ex ante market-based instruments, complementing and enhancing the price and other risk management tools the Government already has within its portfolio. Such a pioneering transaction would be a first for a sovereign entity in Africa and enable Malawi to manage risks, rather than managing crises. Through price discovery it will facilitate Government and donor planning and preparedness for drought-related food shortages and, as an additional instrument, it will assist the Government in developing a comprehensive and cost effective national food security strategy for each agricultural season.

INTEGRATING DISASTER RISK REDUCTION IN DEVELOPMENT

The preceding analysis discussed the relationships between disasters and development. These links provide a framework for conceptualizing the task of disaster management: to reduce the impact of disasters, while promoting development processes that contribute to reducing disaster risks (UNDP 2004). The disaster-development relationship also provides the framework for integrating disaster management in development processes and for transforming disaster management into disaster risk reduction.

Disaster risk reduction is the management of disaster risks to avoid adverse impacts of hazards, through prevention, or to limit those impacts through mitigation and preparedness. The aim is to minimize vulnerabilities and disaster risks through scientifically sound, cost effective and integrated actions that reduce or prevent risks arising from natural and related hazards while taking into account relevant socio-economic, political, cultural and other considerations.

Thematically, disaster risk reduction involves risk awareness, policy and public commitment, and building understanding through knowledge development and information sharing.

Disaster risk reduction also involves adopting the long-term view of reducing prospective as well as currently accumulated risks. Hence, it focuses on proactive management of emerging threats and vulnerabilities, including multiple hazards and vulnerabilities. Therefore, disaster risk reduction is synonymous with disaster risk management. The latter is the process of identifying, evaluating, selecting and implementing actions to reduce risks from disasters to elements at risk.

Addressing disasters as a development issue requires an integrated approach to disaster risk reduction that is mainstreamed in development processes. The UN/ISDR and UNDP have proposed a framework to understand, guide and monitor disaster risk reduction. This framework consists of five thematic areas: (i) political commitment and institutional aspects, (ii) risk identification, (iii) knowledge management, (iv) risk management applications, and (v) preparedness and emergency management. The first thematic area deals with governance issues such as policy and planning, legislation, resources and organizational structures. The second covers risk assessment, impact assessment, and forecasting and early warning. The third (knowledge management) component deals with information management and communications, education and training, public awareness and research. Broad areas of risk management applications proposed comprise environmental and natural resource management, social and economic development practices, and technical measures such as land use planning and standards.

DISASTER MANAGEMENT AT REGIONAL AND SUB-REGIONAL LEVELS

DISASTER MANAGEMENT POLICIES AT THE REGIONAL LEVEL

The issue of disasters is so important that disaster management has become a core area of activity of most regional economic communities (RECs) and other sub-regional organizations (SROs).

The African Union (AU)

The Constitutive Act of the African Union seeks to achieve human security for the peoples of Africa, which includes strengthening resilience to disasters. Consequently, the AU and its predecessor, the Organization of African Unity (OAU), have been concerned with the issue of disaster management and have made efforts to promote risk-sensitive development on the continent, including through support for disaster risk reduction and management activities.

The African Union (AU) and the former Organization of African Unity (OAU) have undertaken some efforts in supporting disaster management in Africa. Its approach has, however, followed the conventional sequencing observed on the continent: its early concern was with reactive emergency response issues. The OAU set up a Special Emergency Assistance Fund (SEAF) for emergency relief and development assistance in 1985. The Fund is managed by a Policy Committee of Ambassadors and administered by the African Development Bank. However, the SEAF did not only focus on relief activities but supported a wide range of interventions deemed to be promoting local livelihood sustenance and protection from damage from disasters and other emergencies. The Fund has successfully supported various initiatives, including early warning, small-scale irrigation, food storage, reforestation, post-emergency reintegration, dam construction and maintenance, emergency preparedness and post-disaster reconstruction in about 33 AU member states to date, as well as land degradation control in the Sahel-Maghreb region.

The activities supported by the Fund revealed an implicit concern of the OAU with localized community-based disaster risk reduction. It also showed the strong link between disaster management and conflict resolution and peace management processes in the overall OAU agenda. The initial orientation of the organization towards emergency management and post-disaster recovery is being explicitly transformed to focus on preparedness for disaster prevention and mitigation. As part of this process, the AU Commission has proposed the establishment of a mitigation and development window under the SEAF. This is pending approval by the EASF Policy Committee of Ambassadors and the consideration of the AU Council of Ministers.

Other programmes of the OAU/AU that contribute to promoting disaster management are those on conflict early warning being instituted by the Conflict Management and Resolution Office, and those on health and nutrition. The OAU also started a programme to develop a regional institute for disaster management training but the status of that initiative is not clear to the study team. In addition, the AU is playing its role in providing policy direction for and popularizing the approach of disaster risk reduction globally and acts as a champion in Africa. The AU is a member of the Inter-Agency Task Force of the UN International Strategy for Disaster Reduction (UN/ISDR).

The AU has focused relatively little attention on engendering political commitment at the national level, developing institutional structures, emphasizing risk identification and promoting knowledge management at all levels in its disaster management activities. However, the previous experience of the AU in disaster management provides a basis for further determining and developing the functions of the AU in disaster risk management.

The experience of the AU with disaster relief and reconstruction, as well as its advisory role in disaster risk reduction in the international arena, have contributed to the momentum within the AU Commission to promote a regional strategy, programme and mechanism for disaster management in Africa. The coming into being of NEPAD provides the opportunity to achieve that objective.

The New Partnership for Africa's Development (NEPAD)

NEPAD is a vision and a strategic framework for Africa's renewal. The AU formulated NEPAD as its programme to help it achieve its development goals. NEPAD represents a framework for partnership among African leaders, between the leaders and the citizenry, and between Africa and the international community to eradicate poverty and ensure sustainable growth in Africa and to assure the continent's greater inclusion and participation in a globalized world. The objective of this initiative is to consolidate democracy and sound economic management on the continent.

Current programmes developed by the NEPAD Secretariat within the main sectoral priority areas of NEPAD contain specific components that directly or indirectly contribute to the reduction of risks from disasters triggered by natural hazards. Implementation programmes in the environment and agriculture sectors contain sections on disaster management while the entire programmes in education, health, regional infrastructure and market access all contribute to strengthening the resilience of African communities.

For example, all programme areas under the Action Plan of the NEPAD Environment Initiative are relevant to reducing disaster risks but the following are central themes in disaster management: assessment and early warning, land degradation and drought, climate change, and water resources (UNEP 2003). And the NEPAD Comprehensive Africa Agriculture Development Programme (CAADP) proposes interventions in the agricultural sector under land and water management as well as in food security that all have direct implications for disaster management. The following are particularly central to a regional disaster management strategy: (a) strategic regional and sub-regional food reserves as emergency buffer stocks, (b) capacity building for forecasting, preventing and mitigating adverse effects of natural disasters, (c) improving water management, including water harvesting, (d) combating desertification, and (e) emergency prevention systems (EMPRES) for pest and diseases

DISASTER MANAGEMENT POLICIES AT THE SUB-REGIONAL LEVEL

Economic Community of West African States (ECOWAS)

ECOWAS has not yet developed a sub-regional disaster management strategy or programme. However, it is developing programmes in environment and natural resource management, covering desertification control, water resource management, floating weeds control and meteorology, all of which have implications for disaster management. A Sub-Regional Action Programme to Combat Desertification in West Africa and Chad (SRAP-West Africa) contains guidelines for desertification control policies, strategies and actions at the sub-regional and national levels. Its eight (8) priority areas include early warning and drought impact alleviation, while two others focus on enhancing sustainable management of shared water, plant and animal resources. The SRAP provides a strategic and programmatic framework for integrating any disaster reduction and management initiative into poverty reduction, environmental protection and sustainable development planning in the sub-region.

The community is also currently developing a sub-regional Common Agriculture Policy (CAP) and a sub-regional programme for food security. The latter will build on and incorporate elements of the sustainable food security strategy for poverty alleviation in the Sahel, a strategy that is being implemented by the Permanent Inter-State Committee for Drought Control in the Sahel (CILSS).

Furthermore, one of the missions of one of the three organs that support programmes under the Protocol Relating to the Mechanism for Conflict Prevention, Management, Resolution, Peace-Keeping and Security is humanitarian intervention in times of disasters. Also the Mechanism has established an observation and monitoring system with zone-based field observatories in four member states. The mechanism provides a potential strategic, communication, information and operational backbone for a sub-regional disaster reduction and management initiative, particularly in: (i) early warning, (ii) disaster management capability development both at the sub-regional and national levels, and (iii) the coordination of national disaster management platforms.

Thus, the SRAP, peace and security programmes, CILSS programmes as well as the agriculture policy and food security initiatives provide a framework for a comprehensive sub-regional disaster management strategy in ECOWAS. The top leadership of the ECOWAS Secretariat recognizes the urgent need for a common sub-regional disaster management strategy and intends to pursue its development. Efforts are underway with the support of NEPAD Secretariat and the UN/ISDR to initiate this process.

Southern African Development Community (SADC)

Disaster management has been a very important component of the overall SADC strategy regional development. The very first SADC Summit recommended the development of a regional disaster management strategy. Consequently, the sub-region has since made significant progress in developing and institutionalizing a sub-regional disaster management strategy.

At the operational level, the Disaster Management Steering Committee provides technical support for sub-regional efforts in disaster management under the leadership of the Council of Ministers. The Regional Disaster Management Unit mobilizes resources and coordinates emergency preparedness and relief and recovery interventions, while a Disaster Response Task Force is activated during regional emergencies such as the recent 2001-2003 food crisis. Under the new institutional restructuring initiative, the Steering Committee is likely to be transformed into a ministerial management oversight committee that reports to the Council of Ministers.

The SADC Sub-Regional Disaster Management Strategy, approved by the SADC Council in August 2001, is facing severe implementations problems. The Council made further recommendations at its August 2002 forum in Lusaka (Zambia) to improve the coordination mechanisms of the Strategy, while the implementation problems were again of concern at the December 2002 meeting in Maseru (Lesotho).

The major thematic or sectoral areas of the SADC disaster management programme cover food security, climate and environment and water management. The SADC sub-regional strategy has established a Regional Vulnerability Assessment Committee (RVAC) to oversee the conduct of vulnerability assessment and food security monitoring through the use of a common methodology and approach by National Vulnerability Assessment Committees (NVACs) and other partners, including relevant non-governmental organizations. This mechanism provides the basis for extending the assessment to cover other vulnerabilities and hazards, such as water and health sectors, as recommended by the SADC Council.

Regarding drought and flood, the SADC Water Sector Coordinating Unit (WSCU), in consultation with the SADC Special Programme responsible for establishing the sub-regional disaster management mechanism, has approved a strategic approach to managing floods and drought. In this regard, the sub-region has made significant strides in developing a common early warning strategy and system for drought, flood and windstorm hazards. At the sub-regional level, the SADC Regional Early Warning Unit (REWU) develops information on weather threats, drought conditions and food security potentials based on two-way flow of data and information between it and National Early Warning Units (NEWUs). The Regional Remote Sensing Unit (RRSU) plays a significant role in this sub-regional early warning system through activities such as receiving and analyzing remote sensing and geographical information systems (GIS) data as input into the sub-regional early warning systems. The REWU also works closely with the African Centre of Meteorological Applications for Development (ACMAD).

The key institutional player in the SADC hydrometeorological hazard early warning system is the SADC Drought Monitoring Centre of Harare (Zimbabwe) that monitors extreme climate events, particularly drought and floods, and provides warning advisories to member states and relevant institutions to aid the development and implementation of appropriate drought mitigation policies and programmes. Its information products include monthly and seasonal climate forecasts, tropical cyclone warnings, decal drought watch information and synoptic climate reviews. Other major institutions involved in hydrological forecasting are the WMO tropical cyclone monitoring centre (WMO-RSMC) in the French island of Reunion that serves the sub-region, especially the members of the South West Indian Cyclone Committee (SWIO) and the Regional Meteorological Specialized Centre.

The two aspects of disaster risk management, proactive and compensatory disaster management, have been split in allocating responsibilities to the various structures of SADC. This is because while the current SADC Disaster Management Framework has strong institutional links with the Food, Agriculture and Natural Resources (FARN) sector, emergency management functions are primarily the responsibilities of the political, defence and security mechanism.

The SADC emergency management programme is being strengthened in line with Article 2 of the Treaty and Protocol on Politics, Defence and Security of SADC which stipulates the overall objectives for the SADC regional disaster management strategy on emergency management. The SADC Secretariat is currently drafting a Strategic Indicative Plan for the SADAC Organ on Politics, Defence and Security with one objective of strengthening regional capacity in disaster management and coordination of international humanitarian assistance. The strategies for achieving this objective are to enhance regional disaster management mechanisms, particularly national capacities in early warning, document regional intervention capacity and coordinate the programmes of the regional committee on natural disasters.

The importance attached by SADC to emergency management within the context of its political agenda shows the strong link between natural disaster management and peace and security within the sub-regional disaster management strategy and, by implication, that of its member states. Also the SADC sub-regional disaster management programme is strongly linked to its broader regionalization agenda because a key objective of the programme is for disaster risk management to actively and synergistically promote regionalism.

The SADC disaster management programme has emphasized political commitment and institutional arrangements in addition to emergency management. Regarding the former, the programme has made progress in developing sub-regional strategies, as in the case of the indicative framework for floods and droughts management. The SADC disaster management programme has adopted participatory approaches and utilizes national structures (such as in early warning and vulnerability assessment) but promotes sub-regional cooperation (such as in river basin management). The programme further emphasizes risk identification by investing in the development of early warning and vulnerability assessment capabilities in the sub-region. However, the SADC disaster management programme has placed relatively less focus on knowledge management activities, although it supports risk reduction research activities of the Drought Monitoring Centre (DMC)-Harare. Also the gender aspects of the sub-regional strategy are not explicit.

Given the above gaps, the SADC disaster management strategy and programme could be improved through:

- Enhancing performance of its institutional structures for disaster Management;
- Increasing the coordination of the components of the strategy, including integration of food aid in overall disaster management strategy;
- Promoting knowledge management, including greater public awareness of disaster risks and disaster management programmes;
- Enforcing compliance with protocols on river basin management; and
- Enhancing its gender-sensitivity.

The Economic Community of Central African States (ECCAS)

Similar to the situation in West Africa, ECCAS has not yet developed a sub-regional disaster management strategy, programme or institutional structure. The Community has launched a renewal programme to achieve its poverty and development goals.

As part of the process of increasing attention to human security risks, ECCAS has established a Department of Humanitarian Affairs aimed at emergency response. This Department could provide ECCAS with the institutional base upon it could develop a comprehensive sub-regional disaster management strategy. ECCAS has also put in place an early warning system. The ECCAS Secretariat has recently been actively participating in regional disaster risk reduction activities, including consultations, and has expressed interest in developing a sub-regional strategy for disaster risk management, with the support of UN/ISDR.

Inter-Governmental Authority on Development (IGAD)

IGAD has developed a regional strategy to strengthen sub-regional disaster preparedness and response capabilities, which incorporates a Regional Programme for Disaster Risk Management. This Programme aims at disaster mitigation through capacity development and enhancement. Specific objectives of the Programme include: (a) facilitating the development and effective implementation of policy and legislation frameworks and programme interventions among member states, (b) strengthening community participation in disaster issues, (c) establishing a sub-regional mechanism, and (d) promoting international cooperation.

The Programme components are: (a) development of disaster risk management strategy and planning process, (b) elaboration of policies, legislation and agreements for disaster management in member states, (c) improvement of regional collaboration in disaster risk management, (d) strengthening of early warning information systems and vulnerability analysis, (e) development of training and education for disaster management, (f) enhancing capabilities for disaster needs and impact assessment as well as resource mobilization, and (g) improving preparedness for relief and rehabilitation assistance management.

The implementation arrangements for the programme will involve a regional mechanism at policy level that will be set up to give political guidance and support to the IGAD Secretariat. In addition, a Regional Disaster Risk Management Technical Advisory Panel, composed of senior technical personnel from IGAD member states, will advise the policy organ while a Programme Management Unit will be responsible for programme implementation.

Meanwhile, as part of ongoing efforts to regionalize disaster management, IGAD operates a Regional Early Warning System (REWS) as a key component of national drought and flood preparedness and response programmes in the Greater Horn of Africa (GHA). The Drought Monitoring Centre in Nairobi closely services this regional system partly by integrating various data inputs

- including its own analytical outputs - in developing operational hydrometeorological information for disaster planning through the consensual mechanism of the GHA Climate Outlook Forum (GHACOF), a mechanism that operates along similar lines as those of SADC and CILSS countries. Under a regional integrated information system, efforts are under way to integrate drought and flood disaster early warning and environment and natural resource information.

In addition, as part of its efforts to promote sustainable natural resource management, IGAD has worked with member states to develop a sub-regional Action Plan to Combat Desertification (SRAP).

Maghreb Arab Union (UMA)

UMA has developed neither a regional disaster management framework/ strategy nor programmes.

Indian Ocean Commission (IOC)

The IOC does not currently have a sub-regional disaster management strategy or a specialized institutional mechanism. However, it has some experience in implementing regional initiatives related to disaster management, through the implementation of specific projects such as the first and current phases of the Regional Cooperation Programme in Meteorology. In cooperation with other regional groupings (COMESA, IGAD and SADC), IOC is also involved in a project called Preparation for the Utilization of the Second Generation Meteostat in Africa (PUMA). This project is under the auspices of the European Union-assisted African Monitoring of the Environment for Sustainable Development (AMESD) initiative. In addition, IOC has developed a Regional Coral Reef Monitoring Network of the Member States of the Indian Ocean Commission, in conjunction with the Global Coral Reef Monitoring Network (GCRMN) and with the support of the World Bank's World Environment Fund.

These project interventions, as well as the positive orientation towards disaster risk reduction in some of the IOC member states such as Madagascar, bode well for the future in terms of developing a common sub-regional comprehensive disaster management policy and institutional framework. IOC has organized a sub-regional consultation on disaster risk management and plans to develop a sub-regional strategy with the strong support of UN/ISDR Africa.

UNCERTAINTY, RISK AND RISK MITIGATION

A farmer needs to make investment every season in working capital, which makes timely credit a necessary ingredient for the success of their production activities. Even after making right investments, a farmer may not get due return because of unforeseen reasons which are beyond the control of the farmer. In other words, farm income being uncertain, appropriate risk mitigation strategies are necessary for smoothening of income of the farmers.

There are mainly three types of risks emanating from three sources of uncertainties. These are:

- Production risks
- Price risks and
- Input risks.

Production risk may arise owing to two factors: Weather risk and risk from pests and diseases. Price related risk arises due to sudden change of demand and instability in expectation formulation. Farm households mainly face the price risk because, production decisions are made far in advance of the date when output is realized. Input risk occurs when there is either a shortage of inputs or when their prices vary.

While all three types of risks appear to be present in the state of Karnataka, production risk arising from uncertain weather is more prominent.

PRODUCTION RISK ARISING FROM UNCERTAIN CLIMATE

As is well known Karnataka is a drought prone state with comparatively low amount of rainfall. 'Two-thirds of Karnataka's geographical area is arid or semi-arid. Out of 27 districts, 18 districts are drought prone with annual normal rainfall of less than 750 mm. The normal annual rainfall of the state is 1,139 mm, received over 55 rainy days'. Irrigation facilities in the state have also not developed adequately to address the problems of the farmers. Our survey includes both irrigated and rain fed areas and there is stark difference in farmers' conditions in these two situations.

It is worth noting that in the recent year state has also seen disaster due to flood. Though the state of Karnataka have an early warning system to understand the climatic aberrations, it would be necessary to translate this into contingency plan to put in action at the instance of the first signs of climatic distress.

In order to work on this, it will be necessary to provide institutional training to the farmers as well as enhance the capability of the farmer to interpret and work on a given operational and decision-making universe.

There is also severe shortage of agriculture extension officers who are supposed to impart technical knowledge to the farmers. Many farmers reported during our survey that they hardly see an extension officer in their village. Further farmers also felt that extension officers themselves lack the knowledge and they should be trained with respect to soil testing, identifying appropriate pesticides etc...

During our survey, we have observed that weather related disaster impacts almost all the farmers. Only 7 per cent of households reported that though they faced adversities, their crops have not been destroyed substantially.

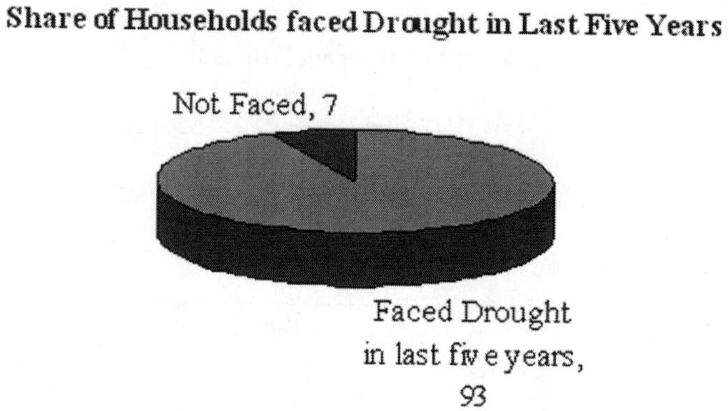

Fig. Share of Households Faced Drought or Flood at Least Once in the Last 5 Years

Average number of drought situation (which had severe impacts on crop) faced in last 5 years by farmer households is about 2. Thus, one can observe that every alternate year one is faced with weather related adversities. Given the intensity with which a farm household is subjected to whether related risk, it is essential to address the issue effectively.

In this regard, lack of irrigation facilities even in the parts of developed districts like Mandya or Chamarajanagar that severely affect the farmers is worth mentioning.

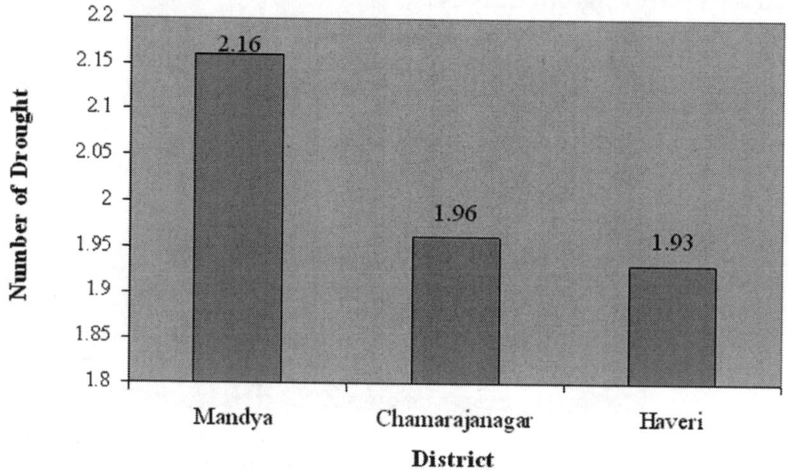

Fig. Average Number of Droughts Faced in the Last Five Years

In case of severe drought in an area, Government declares aid to the farmers. However, even with all good intentions of the Government, such help does not reach the needy. In our sample, though almost all farmers faced weather related disasters but only a few (about 30 per cent) have received Government help.

Fig. Share of Households who got Benefited from Government during Drought

No doubt, natural disaster drastically reduces a farmer's income. Marginal and small farmers usually have little savings. As a result, a large percentage of farmers need to borrow in order to meet their expenses. These borrowings are done often from friends and relatives who are economically better off. Another way of compensating income is to get engaged as labour in agriculture or in other works. Of late, we have also observed farmers with small land holdings are working under NREGA programme. A few farmers who have small business like petty shops fall back on the income earned from such non-farm activities. Apart from this, 15 per cent of farmers reported to have sold assets during our survey.

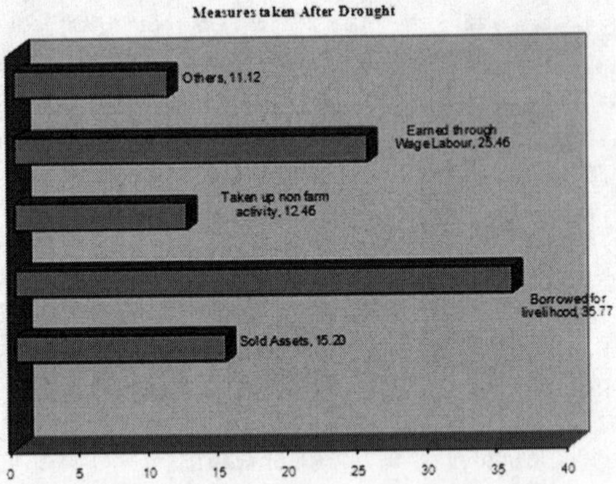

Fig. Means of Survival After Drought

Thus strategies to compensate in case of weather related aberrations is rather weak and no doubt it is necessary to have a systematic approach to the problem. Developing non-farm activities is one such option and problems related to this are discussed in the sequel. In addition, irrigation can go a long way and we see significant difference between the farmers having access to irrigation facilities and those who do not. Risk mitigation strategies like insurance are though essential, at present are underdeveloped and issues pertaining to crop insurance are also taken up.

PRODUCTION RISK ARISING FROM PEST AND DISEASES

The farmers also face pest and disease related problems, which in turn reduces their output. During our survey it has been revealed that Mulberry crop growers, for instance, face disease related problems during winter due to certain worms attacking the plants. Paddy growers face paste related problems. Most commonly faced problem of paddy growers are blast disease, disease from yellow stem borer, disease from brown plant hopper, Gandhi Bug, and Udubatta disease. The cotton producers, on the other hand mainly face problems from borer insects. During our survey, the farmers complained that though they spray insecticides, due to substandard insecticides, insects have become resistant to them and can destroy crop. Rats are also a menace to the farmers. Farmers still not have upgraded their mitigation strategies in this regard.

Price Related Uncertainties

As mentioned, in addition to weather and disease related uncertainties, farmers also face price related uncertainties and we look at this issue as well. The major crops covered in our study area are Paddy, Ragi, Maize, Sugarcane, and Cotton and we have collected the minimum support price of these crops.

Table. Minimum Support Prices Over the Years

Year	Paddy	Ragi	Maize	Sugarcane	Cotton
2000-01	510	445	445	59.5	1625
2001-02	530	485	485	62.05	1675
2002-03	530	485	485	69.5	1675
2003-04	550	505	505	73.5	1725
2004-05	560	515	525	74.5	1760
2005-06	570	525	540	79.5	1760
2006-07	580	540	540	80.25	1770
2007-08	645	600	620	81.18	1800
2008-09	850	915	840	81.18	2500

It is seen that the increase in minimum support price had been minimal over the years. In fact, the jump in minimum support price is seen only during 2008-09. Another information revealed during our survey is that while government declares minimum support price, there is little procurement of crop at that price. Though many farmers wish to sell to the state agencies at the minimum support price, they are unable to do so due to lack of demand. Thus, they feel that minimum support prices are not of much use to them as procurement falls far below their supply. The farmers often need to sell at market prices. Therefore, we have also examined the market prices of these crops. It is observed that market prices of crops have fluctuated over the years. This clearly shows the kind of market situations and uncertainties faced by the farmers. However, data collected from farmers in the recent year have shown increase in market prices of all crops. This may corroborate the sharp increase in food price inflation during the last two years.

Table. Prices of Different Crops Across the Selected Districts at Different Points of Time

Year	Paddy	Ragi	Maize	Sugarcane	Cotton		
	Mandya	Haveri	Mandya	Chamarajanagar	Haveri	Mandya	Haveri
2006-07	660	626	663	944	625	900	2188
2005-06	687	577	475	540	745	1100	1865
2004-05	637	697	401	659	688	-	2075
2003-04	557	706	444	500	492	-	2324
mean	635	651	496	661	637	-	2113

Note: Prices for all crops are not available for all districts as these crops are not grown

Table. Summary Measures of Prices from Primary Survey 2009-10

	Paddy	Ragi	Sugar	Maize	Mulbery	Cotton
Mean	877	883	1217	766	123	2514
Median	850	800	1100	770	120	2500
Mode	900	800	1100	800	120	2200

INPUT RISKS

In addition to production and price risks farmers also suffer from input related risk.

During our survey almost all the farmers informed that they face shortages of seeds and fertilizers. There also exists problem of timely supply of the inputs, which in turn also reduce production. Most of the farmers avail seeds and fertilizers from Raita Samparka Kendras. But quantity supplied is not sufficient. Hence they also depend on private traders who often provide substandard inputs at higher prices.

RISK MITIGATION: CROP INSURANCE

One of the standard ways of mitigating risk is through crop insurance. Unfortunately, however, use of crop insurance is not wide spread in Karnataka. In our sample households, only 17 per cent of the households are covered by insurance scheme and the rests are totally uncovered. Thus, in case of any calamities they are driven to a distress situation.

One of the major reasons for not taking up crop insurance is lack of awareness among farmers and absence of insurance scheme in the regions. Thus we feel that formulation of appropriate crop insurance schemes as per need of specific regions is of significant importance.

We also notice from our sample that around 40 per cent of the farmers are not interested in crop insurance. This is because crop insurance is usually area based.

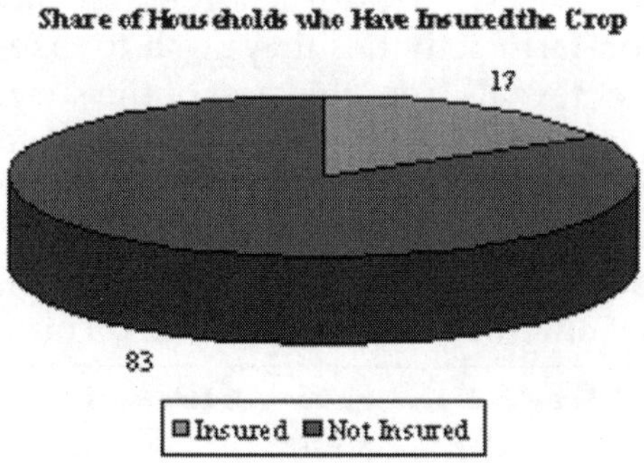

Fig. Share of Households with Crop Insurance

The "area approach" is operated under the results of crop-cutting experiments. Each year a set number of plots with the insured crops for a certain "area" are used as the indicators of an individual farmer's losses within that area. Insured farmers receive indemnity based upon the difference between the threshold yield and the yield of the crop-cutting experiments in their area. Crop yields naturally vary even over small areas and very localized natural calamities could occur. Situations easily exist in which farmers would not be compensated for their loss.

Therefore a farmer is wary about getting compensation. A farmer feels that he/she may be paying insurance premium unnecessarily. If there is a large scale calamity he would, in most cases get government compensation without any insurance cover. Thus, the farmer considers an individual land based insurance as the best option. However, such option has premium implications which they fail to grasp. Given this scenario, it is necessary to take the farmers into confidence while formulating the insurance policies. What can be the premium related implications of an individual based insurance should also be discussed; otherwise, the entire exercise is going to be futile.

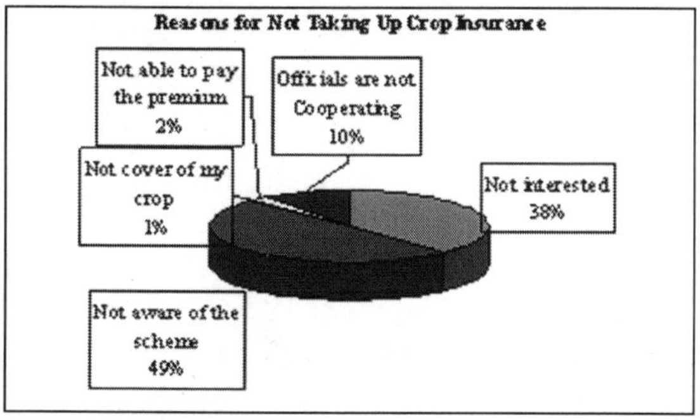

Fig. Reasons for not Taking up Crop Insurance

NON FARM ACTIVITIES

Once a farmer faces a situation of depressed income, he can at least get over the distress condition, in case he has nonfarm activities. Most of the farmers in our sample have no non farm activities. We have therefore enquired about the possible non-farm activities they wish to take up.

An interesting fact reveals from this part of the survey. The farmers have no knowledge and imagination about diverse non-farm activities. Most farmers consider certain petty business like opening a small shop or buying a capital good such as tractor or truck for renting is their only option. Thus there is a need to provide training to the farmers as per the resource base of a region to develop meaningful non farm activities. This can go a long way.

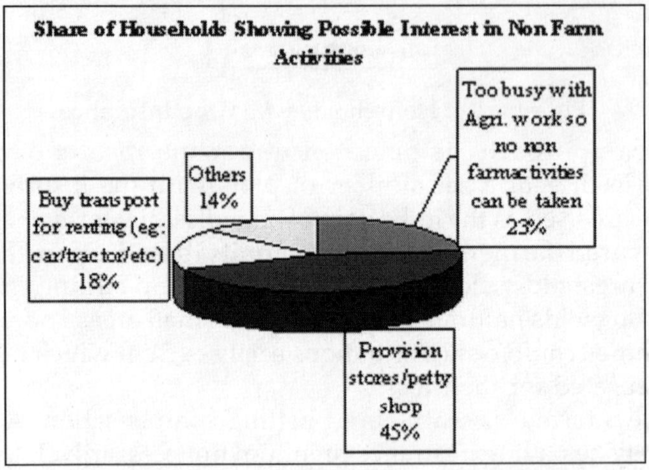

Fig. Share of Households Showing Interest in Different Types of Nonfarm Activities

Currently most farmers have not taken up non-farm activities due to lack of information and resources. It is rather difficult to get credit from formal sector for such activities without a proper project plan.

3

Drought Mitigation Interventions in the Pastoral Livestock Sector

The interventions concerned had already been evaluated by Aklilu and Wekesa (2001) and the authors are indebted to their work. Our original intention was not only to complement their analysis with more in-depth participatory evaluations, but also a more formal cost-benefit analysis. Despite efforts to collect quantified data from implementing agencies, this second aim was only partially successful, for reasons that throw light on the need for more co-ordinated monitoring and evaluation of drought mitigation. This report will present: firstly some elements of a comparative cost: benefit analysis and lessons therefrom; secondly the most important cross-cutting lessons from the participatory evaluations; and thirdly lessons learnt about each of the major types of drought mitigation intervention.

CROSS-CUTTING ISSUES FROM THE PARTICIPATORY EVALUATIONS

Four projects were subject to participatory evaluation using semi-structured interviews and established PRA tools. In Marsabit District, interviews were held in seven locations over two weeks, and in Samburu in three locations over one week. Separate group interviews were held for male and female pastoralists. The participatory evaluations focused both on the processes of drought mitigation, and in particular the extent to which there had been meaningful community participation in planning interventions, and the impact of interventions on different strata, including those without livestock.

Table. Projects Subject to Participatory Evaluation

Marsabit District	Samburu District
• ACK Destocking Programmes	• COOPI/Ramati Development Initiative Emergency Veterinary Programme
• ACK Programmes of Supplementary Feeding of Livestock	
• COOPI/CIFA Emergency Veterinary Programmes	

Generally, the evaluations revealed a greater dissatisfaction about the processes of intervention than was portrayed by Aklilu and Wekesa. Pastoralists felt that the interventions had either been implemented without consultation, or that the committees formed had not functioned (in three of the four communities benefiting from destocking). They would have preferred elected community committees, an element of training for community members, and more integration of emergency projects with long-term development. Men were better informed than women about the projects, and women seem to have been a lot less actively involved in implementing the projects.

In terms of impact, there was a concern that the interventions evaluated had not helped households with small livestock holdings, who would need alternative forms of assistance such as income-generation projects. There was even a suggestion that by insisting on the exchange of livestock for services or cash, some projects may have reduced food security for the poorest households.

In comparing the different interventions, pastoralists felt that destocking projects maintained their purchasing power during the drought. Supplementary feeding of livestock and veterinary interventions were more important in allowing rapid recovery after drought.

FINDINGS ON SPECIFIC DROUGHT MITIGATION INTERVENTIONS

The most important drought mitigation intervention tested so far is emergency livestock purchase. One form of this is where the implementing agency directly buys weak animals at above the market price. Animals are usually slaughtered and the meat distributed locally to schools, hospitals, orphanages or poor households. Alternatively, transport subsidies are offered to traders to encourage them to purchase stock thereby strengthening livestock markets. Destocking has several purposes. It allows pastoral households to liquidate some of their capital assets (livestock) before they are lost and increases the purchasing power of these households. Some value is therefore salvaged from animals, which may otherwise have died and meat or stock can be redistributed to needy households. In theory protecting herders' purchasing power by buying animals where markets are absent creates two categories of beneficiary: those who sell animals to the intervention/project at subsidised prices, and those who benefit from the general rise in prices on local markets caused by the extra demand created by the intervention/project.

Destocking can also:

- create a market for weaker animals, thereby enabling herders to keep stronger animals in their herd, preserving a key household capital asset for post-drought recovery
- Impact on the nutritional status of poor households and contribute to school and other feeding programmes

- Support the trading activities of women's groups
- Reduce overstocking around village settlements.

A number of small agencies solicited funds for destocking in northern Kenya during the 1999-2001 drought. For example the Anglican Church of Kenya (ACK) used donor funds to purchase over 6000 sheep and goats from local pastoral households in Marsabit District in exchange for cash and high protein energy feed. Purchased animals were slaughtered and the meat distributed to needy households. Some animals were retained for restocking and 30 animals were given to households who had lost all their animals during the drought. Pastoral household purchasing power was maintained allowing beneficiaries to purchase feed and drugs to maintain their remaining animals, buy food for the household and pay school fees. There is some evidence that the purchase programme also stabilised livestock market prices.

VSF-Belgium facilitated the purchase of 13,000 smallstock by women's and youth groups in Turkana District. The groups purchased the animals themselves and the agency bought the dried meat from them for distribution in schools and hospitals. Many of these animals would not have been slaughtered without the intervention of VSF. The purchasing power of livestock owners was improved, the groups slaughtering animals increased their incomes and the nutrition of hospital patients and school children improved.

ALDEF assisted poor women's groups with micro-credit to purchase more than 9,000 smallstock, 95 cattle and 194 camels in Wajir District. ALDEF purchased the meat from the groups for redistribution to needy households and organisations. Meat was made available to 17,000 beneficiaries (mostly urban poor) and 7000 pastoral households improved their purchasing power. The total cost of this intervention was $US193,000 and total benefits (cash transferred to pastoral households for animals purchased and the value of the meat distributed) were estimated to be $US260,000.

As animal condition declines during drought, livestock traders become reluctant to risk purchasing animals for which there may be limited demand in terminal markets (e.g. Nairobi). A transport subsidy had been successfully used in Isiolo District during an earlier drought to assist pastoral households market stock and resulted in increased offtake as traders risks were reduced (Barton and Morton 2001, Morton and Barton forthcoming). It also increased pastoral household purchasing power as many of the beneficiaries inhabited remote areas and would not have been able to market stock without this intervention (it was not possible to trek animals to market as water and pasture were scare on stock routes).

NORDA, an NGO based in Mandera District offered traders a subsidy to the value of one third the cost of transport (by truck) to Nairobi. Subsidies provided were for a total of 22,000 sheep and goats. It is not clear however, how many of these smallstock would have been marketed without the subsidy.

A similar operation in Turkana (VSF-Belgium) provided subsidies for both movement to terminal markets and movements within the district but was beset by fraudulent claims for subsidy.

It is clear that the modus operandi for transport subsidies should be carefully planned to avoid fraud. A transport subsidy might however be more cost effective than a destocking/animal purchase programme in very remote locations. The end result from the pastoral household point of view is the same (i.e. improved purchasing power and salvaging of some capital). A subsidy should therefore be targeted at those locations not normally visited by traders. Under these conditions it is much easier to assess impact, where marketing normally takes place a subsidy may not be required as many of the animals purchased may have been bought and sold without the subsidy.

Veterinary interventions

During drought as animals weaken in the face of pasture and water shortages they become more susceptible to disease and parasitic loads. The objective of veterinary interventions is therefore to assist pastoral communities combat these diseases, particularly in their most valuable animals (breeding stock and loading camels). Improved survival of breeding stock will allow the rapid recovery of herds and livelihoods post drought.

For example COOPI/RAMATI and CIFA provided veterinary assistance in three Districts, Samburu, Marsabit and Moyale. Payment for drugs was in the form of goats, which were slaughtered, and dried meat distributed to schools to strengthen the school-feeding programme during the drought. Between 5-6,000 households benefited from animal health provision on a monthly basis and over 40,000 smallstock, 2,500 cattle, 3,700 camels and 72 donkeys were treated. Services were delivered by 100 community animal health workers (CAHWs) employed by the implementing agencies. Although the number of animals saved by this intervention is not known it was estimated that it may have contributed to the survival of 20 per cent of the animals treated.

VSF-Belgium working with local communities in Turkana facilitated the treatment of 74,000 animals and the vaccination of 97,000 smallstock against Contagious Caprine Pleuropneumonia (CCPP) on a cost recovery basis. The vaccination was designed to assist the recovery of the livestock population post-drought as animals become very susceptible to CCPP when the rains return. It is not known how many animals were saved by this intervention.

Supplementary feeding

If pastoral households lose valuable breeding stock during drought it can take many years for their herd to recover. There is a danger that poorer households will never be able to rebuild their herds and as a consequence fall out of pastoral production. It is also important for pastoral households to maintain some stock to be able to take advantage of the, often good, grazing conditions which follow a drought. There is widespread anecdotal evidence that pastoralists sometimes use relief grain intended for human consumption to feed their herds.

ACK-Marsabit integrated a supplementary feeding intervention with their de-stocking activities. Feed was exchanged for livestock which were slaughtered and the meat distributed to needy households. 180 tonnes of high energy, fat and protein concentrate feeds was purchased which was sufficient to maintain 8,000 smallstock for a period of 3 months. Total costs of this intervention were USD 48,000 and benefits were estimated to amount to USD128,000.

This sort of linkage of supplementary feeding to destocking may help avoid one of the major drawbacks of supplementary livestock feeding when it is scaled-up: that it artificially maintains herd levels and thus exacerbates environmental degradation. This has been a controversial issue in Middle Eastern and North African countries (with middle or high per capita average incomes) where feed distribution during drought is practised massively and increasingly delinked from objective meteorological drought. There is little likelihood of either donor or domestic funds being available in Kenya for intervention on this scale, but the environmental implications of maintaining herd numbers through drought, and selective procedures to mitigate them, must be borne in mind.

Water Provision

Assistance with the provision of water for humans and livestock includes borehole maintenance as well as the drilling of emergency and contingency boreholes. An example is given by the drilling of the emergency borehole at Harakhotkhot in Wajir in 2000 by Oxfam. The emergency borehole enabled fifty families, previously using another borehole 70 km away, to water their animals locally. Estimated reduced mortality from the reduction in watering stress was valued at $US 64,300, compared to a construction cost of the borehole of $US 38,000. This does not take into account considerable additional benefits, including reduced animal mortality in later years, and reduction in women's time spent fetching domestic water. Provision of boreholes approximates to a public good.

Borehole repair and maintenance shows similar positive returns. Analysis of borehole maintenance in Wajir suggests that each unit of Ksh 1 million ($US 13,300) spent on maintenance or rapid repair of a particular borehole will be justified if it substantially reduces the risk that 125 cattle (estimated price Ksh 8,000 each), or 1,250 sheep or goats, or some combination of these, will die following borehole failure. Given that in a drought the average borehole is serving several thousand cattle and several tens of thousand of sheep and goats, these are plausible assumptions. As before, this ignores further benefits to women in terms of reduced time spent collecting domestic water. Borehole repairs and maintenance may be considered a private good. If so, micro-finance products could be designed to allow pastoralists to contribute to the cost.

In addition to saving animal lives, secure water provision can reduce labour time spent watering animals and walking animals between water and pasture, as well as increasing the quality of pasture available. We were unable to find examples to quantify this. These are private goods.

Emergency water interventions such as borehole maintenance and repair, contingency borehole drilling, and water tankering have a further important benefit in terms of women's time spent fetching domestic water. In conditions of water shortage, women spend many hours each day collecting water, with negative consequences for their own health, income earning opportunities, household survival and child care. Analysis of the water tankering around Wajir town suggests that six months tankering to a particular site, at a cost of USD 7,700, resulted in savings in woman's time fetching water of USD 27,600, in addition to reduced animal mortality valued at USD 12,300.

The true success of contingency boreholes may not be known for some years. Future management of these resources will be crucial before they can be claimed to be a success. In areas where drought-time grazing was opened up by the provision of water it will be essential that boreholes are closed (capped) during periods of average rainfall otherwise environmental degradation is likely to occur, such as that around boreholes elsewhere in northern Kenya. There is a danger that these areas of drought time grazing can become an open access resource if water is available year-round. To date it appears that communities have agreed to close contingency boreholes and to reserve use for future emergencies (for a Kenyan example see Meigh, Robins and Calow 1999, Box 18). Whether this commitment will be sustainable in the longer term is unclear.

Free generator sets and borehole equipment were donated to community groups in several arid districts. This is an example of donor activity distorting local markets for fuel and spares that could lead to further dependency of pastoral communities and is counter to the principle of cost recovery pursued by other interventions. It could be argued that good drought preparedness on the part of communities, donors and local governments would include functioning boreholes with spares and fuel in place ready for any emergency.

Movement of Livestock and People

Movement is critical to pastoral household survival during drought. Allowing some movement of people and livestock across international borders can yield important benefits for pastoral communities, particularly when drought does not affect a whole region. Mobility is an important indigenous coping strategy and should be integrated with other mitigation interventions.

Cross-border and other peace initiatives are therefore an important contributor to the mitigation of drought impacts.

OAU-IBAR had been working in border areas of NE Uganda, SE Sudan, SW Ethiopia, and NW Kenya as part of the PARC programme for a number of years and therefore had established contact with drought-affected communities in Kenya and those pastoral groups less affected by drought in Uganda and Ethiopia. The objective was to use animal health to facilitate peace and reconciliation meetings between different antagonistic pastoral communities (ethnic groups) in order to create an environment conducive to the development of the livestock sector. The purpose of the meetings was to use animal health as an entry point to begin discussions about decreasing raiding and banditry among the Karamojong and Turkana tribal groups and to improve access to critical water and pasture resources.

The project facilitated meetings between elders of the respective groups by arranging transport, food and accommodation. The elders would be allowed to discuss the issues of livestock keeping under one roof for a period of two to three days without the interference of politicians or civil servants, who could have influenced the outcome.

Accessibility to critical water and pasture resources for Kenyan pastoralists was facilitated by this project. It is estimated that nearly 100,000 cattle from Turkana were in Uganda at the height of the drought in Kenya. The cost of the project was US$ 72,000 and it was conservatively estimated that this led to savings of US$ 93,000 in terms of livestock that may have otherwise died due to drought. This estimate takes no account of the value of likely future trade and movement in livestock or of the under-utilised natural resources used by moving livestock and people.

FINDINGS AND LESSONS FROM THE COST-BENEFIT ANALYSIS

For the following projects, data additional to that given by Aklilu and Wekesa (2001) was obtained from the implementing agencies and others and used for a summary cost-benefit analysis. The costs and benefits of projects in each intervention type were aggregated for the analysis.

Table. Drought Mitigation Interventions Targeted for Cost:Benefit Analysis

Type of intervention	District	Donor/Implementing Agency
Destocking	Mandera	NORDA
	Narok	World Concern
	Garissa	CARE Kenya
	Marsabit	CEC
	Wajir	ALDEF
Transport Subsidy	Mandera	NORDA
	Turkana	VSF Belgium
Veterinary	Samburu, Marsabit, Moyale	COOPI

	Marsabit	ITDG/Lutheran World Relief
	Mandera, Garissa, Wajir	VSF Switzerland
Supplementary Feeding	Marsabit	ACK
Conflict Resolution	Turkana	OAU-IBAR
Water	Wajir	Oxfam

a) The terminology follows Aklilu and Wekesa (2001). "Destocking" refers to the purchase of animals by an external agency which then distributes the meat as relief food. "Transport Subsidy" refers to the provision by external agencies of additional incentives to private traders to buy livestock - which was referred to as destocking in earlier work of two of the current authors (Barton and Morton 2001, Morton and Barton forthcoming.

Even after collection of supplementary information, quantitative data on costs and benefits remained very partial, and required a number of assumptions. The data as obtained allowed a summary comparison.

Table. Summary Cost:Benefit Analyses of Drought Mitigation Interventions

Type of Intervention tankering	Destocking	Transport subsidy	Veterinary	Supplement -ary Feeding	Conflict Resolution	Water
Total value of meat distributed	738,515		2,484			
Total cash received by pastoral households	632,485					
Total value of animals saved			8,092,583	74,667	93,333	217,867
Total value of liveweight gain				53,333		
Total value of extra animals marketed		84,758				
Other benefits	1,040					217,200
Total cost	863,407	97,253	1,515,507	48,000	72,646	168,000
Total benefits	1,372,040	84,758	8,095,067	128,000	933,333	435,067
Benefit: cost ratio		1.59	0.87	5.34	2.67	1.28 2.59

*Other benefits were sale of hides in the case of destocking, and opportunity cost of water collection in the case of water tankering.

As can be seen, all the intervention types had positive benefit:cost ratios, with the exception of transport subsidies. The very positive ratio for the three veterinary interventions is however of greater interest.

The result concerning transport subsidies is surprising. Two of the current authors (Barton and Morton 2001, Morton and Barton forthcoming) studied an earlier transport subsidy scheme, and reported not only general satisfaction among all stakeholders but also an, admittedly unquantified, positive benefit:cost ratio.

The difficulties encountered in performing cost:benefit analyses point to the need for co-ordinated monitoring and evaluation of drought mitigation interventions. If government, donors and implementing agencies are going to learn the lessons of drought mitigation, both to choose between types of intervention and to improve the implementation of each type, there will need to be standardisation of monitoring indicators of both costs and benefits, and of the way they are collected.

To be effective, monitoring and evaluation procedures will need to be integral to drought preparedness strategies, so that they can be launched simultaneously with the interventions and with minimum disruption. They should also be transparent so that those who have to implement them can see their worth.

A first step towards such an approach will be to clarify logically the sorts of costs and benefits we can expect from different types of drought intervention. A first attempt at such a table, for four of the intervention types, is given in Table.

A final issue with regard to cost:benefit analysis is that of upscaling. We strongly suspect that the relation of costs to benefits (such as number of animals saved) over different scales of intervention is not linear.

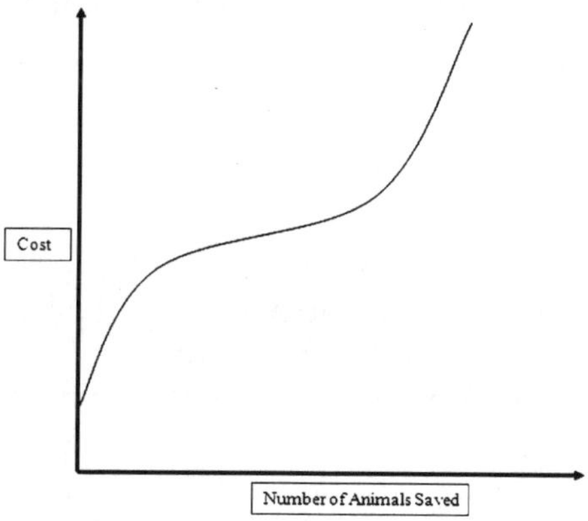

Fig. Likely Relationship Between Cost and Total Number of Animals Saved

IDENTIFY THE CRITICAL SECTORS IN EACH STATE THAT ARE DROUGHT VULNERABLE

In each state, we try to identify those sectors that are major water users, and also important to the state economy. There should be evidence indicating critical impacts of drought on those sectors.

NEBRASKA'S AGRICULTURE

Agriculture is the largest freshwater user in Nebraska. Irrigated land totals up to 7,625,170 acres (2002 census), ranking #2 in the western states (next to California). According to a USGS 2000 water use report, irrigation alone accounts for 72 per cent of total freshwater consumption in Nebraska, with 84 per cent from groundwater and another 16 per cent from surface water sources. Nebraska also has a large amount of dryland crop production (14,895,704 acres) as well as pastureland and rangeland (21,940,679 acres), which heavily rely on water from precipitation. Agriculture is important for the state economy of Nebraska. According BEA 2004 data, agricultural production accounts for 6.4 per cent of Gross State Product; and 1.1 per cent of total employment is in the agricultural sector. Agriculture is important to ensuring a safe and consistent food supply at low prices. Agriculture also has multiplier effect on the state economy through its links to the upstream and down stream industries, including seed firms, fertilizer firms, producers of farm equipment, transportation companies, etc., which makes the total impact of agriculture even greater in the state economy.

Water, as the most critical input in agriculture production, is often the limiting factor in the semi-arid High Plains. A drought in the growing season causes damage to crops and pasturelands, resulting in a loss of production, additional feed costs, and reduced farm incomes. The 2002 drought alone caused a total crop indemnity payment of $290 million (USDA/RMA) within the state. Since the impacts of drought on crop production and on livestock are very different, there is a clear need to separate these effects in our study.

CROP PRODUCTION

Drought has different impacts on dryland production and irrigated crop production. Dryland production relies on precipitation. Reduced precipitation during the growing season would cause low yield, poor quality, and increase abandoned acres. Drought impacts on irrigated crop production are different depending on the irrigation water sources. In Nebraska, most of the irrigation water is from the groundwater aquifer, which is less affected by drought. However, reduced precipitation increases the net irrigation requirement, thus resulting in higher pumping costs.

The surface water availability depends on inflows and reservoir levels, which is more sensitive to drought. For example, the current multiple-year drought and low inflows have resulted in a historical low level of water stored in Lake McConaughy, Nebraska's largest reservoir, during 2004-2006. It was reported that for the first time, the Central Nebraska Public Power and Irrigation District (CNPPID) irrigators received less than a full supply of water in the 65-year history of irrigation from Lake McConaughy. Other major surface water irrigation districts affected by drought are located within the North Platte River Basin and Republican River Basin. The major source of irrigation water in the North Platte Basin is snow in Wyoming and Colorado. While, the other surface water projects are largely dependent on relatively local run-off from rainfall, or on base flow from groundwater. Therefore, drought impacts on irrigation water availability differ by the location and timing of drought.

Livestock

Livestock products account for about 2/3 of Nebraska's farm income and beef cattle are the most important source. Nebraska ranks #2 among the states in total cattle inventory. Drought stressed pastures and rangelands cause poor grass growth. Producers have to pay high prices for supplement feed. A hay shortage in the local market would boost up hay prices. Poor pasture conditions and high feed costs force early marketing and culling, which would damp the market and result in low prices received by farmers. Because of the biological reproduction features of livestock industry, the impacts of drought would last for several years. Extra liquidation would hinder the rebuilding of cow herds and interrupt the normal cattle inventory cycle. For example, U.S. national cattle inventory data from USDA/NASS showed that the 2002 drought extended the otherwise 10-year cattle cycle to 14 years (1990-2004).

ENERGY

Hydropower Electricity

In Nebraska, the net electricity generation from hydropower averaged around 1.3 million Megawatt hours, accounting for only 0.2 per cent of total hydroelectricity generation in U.S. The hydroelectricity generation data from 1990 through 2005 are plotted in Figure. Since 2001, the net hydroelectricity generation has been much reduced, and it might be due to drought.

Thermoelectricity

In Nebraska, thermoelectric plants use more surface water than irrigation (65 per cent of total surface water). The water provides cooling for the plants. So, when the supply of water is limited, cooling costs go up and water reallocation has occurred as needed to meet electrical demand. Another concern with drought is the river's temperature. Users of water must return it at about the same temperature it left the river, so it does not disrupt the ecology of the water plants and animals. So, when the river is already warm in drought, the power plants may have to cut back their generation.

RECREATION AND TOURISM

As there is no ski or rafting sites in Nebraska, the most water related recreation activities are fishing and boating; and the drought impacts are limited. The plots Nebraska State park visits and revenue data from 1980 to

2003 (source: USDOC, Bureau of the Census). We can hardly detect the impacts of drought. Most recreation users and visitors are in-state residents, who spend relatively little money on lodging and restaurants. So, the indirect impacts on the local economy are also small.

SUSTAINABLE LIVELIHOODS FOR POVERTY AND DISASTER RISK REDUCTION

Africa is the only continent whose share of reported disasters has increased over the past decade. Hydro-meteorological hazards (drought, flood, windstorms, particularly tropical cyclones, landslides and wildfire), occur most pervasively and account for most of the people affected by disasters. Africa is particularly prone to drought and flood episodes that are associated with the El Niño-Southern Oscillation (ENSO) phenomena. As illustrated in Figure, droughts are by far the most common trigger of disasters on the continent, and have affected millions of people in the last three decades. These have resulted in the loss of lives and livelihoods, the erosion of assets, mass migration and conflict (Department of Economic and Social Affaires, 2008). Flooding is also frequent, as illustrated by the flooding in Central and East Africa in 2007 and 2008. Together, drought and floods account for 80 per cent of loss of life and 70 per cent of economic losses linked to natural hazards in Sub-Saharan Africa. Famine has also affected millions in East, West and to a lesser extent, Central Africa, with the Sahelian and Ethiopian famines of the 1980s amongst the most publicised of these. While the causes of famines and lesser food shortages are complex, drought, flooding and conflict have often served as catalysts for acute food insecurity, such as seen in Southern Africa in 2001/2002.

Africa is the poorest region in the world. While tremendous developmental gains have been made in the last 50 years, half of Africa's population live on less than one dollar a day and life expectancy is falling in many countries as HIV/AIDS erodes the progress that has been made. These very high levels of poverty increase both exposure to hazards and the vulnerability of the poor to their effects, while hazards in turn diminish and retard progress in improving the lives of the poor and erode their capacity to prevent, respond to and recover from disasters, deepening poverty and adding immeasurably to the numerous challenges Africa faces in trying to achieve the Millennium Development Goals (MDGs). This poverty-risk nexus is thus of critical importance in Africa, as unless this cycle is broken it will be almost impossible for Africans to escape the poverty trap in which many now live. Livelihoods are an essential and illustrative part of this dynamic; sustainable livelihood strategies represent the opportunity to both reduce exposure and build resilience to hazards, but the livelihood lens also provides a window on the accumulation of risk, as the breakdown of livelihoods is indicative increasing vulnerability.

In examining these relationships, it is important to be mindful of the nature of the risk environment in Africa. While large hazard events significantly undermine developmental gains in the region – it is estimated that the flooding in Mozambique in 2000, for instance, lowered the country's Gross Domestic Product (GDP) by 12 per cent - it is relatively silent, recurrent, localised hazards such as recurrent drought, small-scale flooding, inadequate service provision and disease that encroach most on lives and livelihoods. Although seldom measured by conventional disaster-loss tracking systems, these result in incalculable human, crop, livestock, and environmental losses, eroding people's capacities to withstand and recover from hazards, and progressively increase their social, economic, and environmental vulnerability to even modest events.

A range of factors operating at the macro, meso and micro-level shape and influence both livelihoods and disaster risk. Features of the global macro-economic and trade environment have influenced, and continue to effect, investment in the agricultural sector and urban and rural people's access to markets. Structural adjustment programmes have impacted on extent of service provision to the poor and the levels livelihood support provided by governments. Both limited resources and often predatory civil administrations have also influenced the reach and effectiveness of the state, and the extent to which governance supports or hinders people's efforts to earn a living. Communicable diseases such as tuberculosis, malaria and HIV/AIDS in particular, are also undermining livelihoods at a range levels, both directly and through their broader impact on development indicators and governance. Land degradation and desertification also exert a profound influence on livelihoods, as do conflicts. Climate change expected to compound many of these preceding risks and create new ones, suggesting that without more concerted efforts to slow global warming, the livelihoods of millions of Africans will be fundamentally undermined.

The cause and effect of these factors are complex, multi-layered and bi-directional, often simultaneously shaping and being shaped by others. It is impossible to capture the full complexity of situation in Africa in a limited forum such as this, but three distinguishing features of the African experience are:

- The predominance and vulnerability of the rural, particularly small-scale subsistence agricultural sector to drought and flooding;
- The importance of conflict as both a consequence and driver of risk, particularly resource-stressed rural areas; and
- The risks surrounding unparalleled levels of urbanisation on the continent.

This paper examines these issues through three case studies: the regional food insecurity experienced by southern Africa in 2000/2001, the ongoing conflict in Sudan, and the recurrent flooding experienced in poor urban settlements in South Africa, and the Western Cape in particular. It begins by briefly discussing key livelihoods trends on the continent, and the major factors impacting on livelihoods and risk. It then discusses in turn each of the studies, and concludes by briefly drawing out some of the key issues and conclusions suggested by these examples.

LIVELIHOODS IN AFRICA

Relatively few of Africa's poor derive their income from a single source, or hold all their wealth in the form of just one asset. Livelihood studies during the last decade show that increasing numbers of people opt for strategies characterised by multi-tasking and income diversification. Such diversification is pervasive and enduring, and is a common feature of both urban and rural livelihoods. The nature of diversification can vary widely, according to who undertakes it. For higher-income groups it is often an accumulation strategy aiming at maximising profits by investing across sectors, but for the poor it is often a survival strategy, a way of minimising risk minimisation and stabilising income, and usually involves low-skilled, low-paid and often temporary employment.

Most livelihoods strategies on the continent, however, rely heavily on agriculture. The majority of the continent's poor live in rural areas, and is estimated that 70 per cent to 80 per cent of Africans depend on agriculture for consumption and as a source of income – although the proportion of the population engaged in agriculture in sub-Saharan Africa is shrinking rapidly. The sector is an important source of work.

The International Labour Organisation (ILO) estimates, for instance, that 65 per cent of all of the employed in sub-Saharan Africa - 192 million people - work in agriculture. In many cases, the rural poor also depend wholly or in part on natural resources, such as fish, wood and thatch, and in many areas wild foods represent an important food source where other livelihood sources fail. Dependence on the agricultural sector is not confined to rural areas, however, and many of those living in urban areas also derive food and income from agriculture.

Waged work and self-employment is also an important source of income in Africa, particularly work in the informal sector. The last two decades has seen declining or stagnating levels of employment growth in the formal sector in many African countries, and the concomitant expansion of the informal sector, which now dominates the continent's economy both in terms of output and employment. Although it is difficult to measure the size of the informal economy, it is estimated to account for as much 72 per cent of non-agriculture employment in sub-Saharan Africa, and as much as 78 per cent if South Africa is excluded. Such work most often comprises petty trade, and to a lesser extent transport and manufacturing activities, which are often subject to high levels of competition and are poorly paid.

Both rural and urban livelihood strategies in Africa are marked by high levels of mobility, which have been at least partly enabled by improvements in communications and transport technology. Migration has long been a feature of life in Africa, particularly southern Africa, but the expansion of the global village now means that even larger numbers of people migrate from rural to urban areas, and in some countries between rural locations, in search of work. In some regions there is considerable transnational migration, with large numbers of people living temporarily or permanently away from their country of origin.

Growing numbers of people also live on the edge of urban and rural life, commuting from the countryside to the urban centres. Rural-urban migration is contributing to the extraordinarily rapid urbanisation currently away in many African countries. The Africa Commission established by Tony Blair to assess the reasons for Africa's developmental challenges notes that Africa is the fastest urbanising continent in the world. It is urbanising around twice as fast as Latin America and Asia. In 25 years half the entire population will live in cities.

But, while Africa is well on the way to European levels of urbanisation, it does not have the economic base to sustain them. There are no industries to provide jobs and many people – around 72 per cent of the total urban population - live in slums, where they are threatened by eviction, overcrowding and the lack of access to water, sanitation and other services. In the context of weak governance, city authorities also seldom have the funds or the necessary professional staff to manage the rapid urbanisation process. In addition to the direct threat the well-being and security of those living in sprawling slum areas, poorly planned and managed urban expansion also increases their exposure to fires, flooding, landslides and a range of other hazards, and destroys important environmental services.

KEY FACTORS IMPACTING ON LIVELIHOODS IN AFRICA

Libaralisation, Trade and Globalisation

The last few decades have seen the Africanisation of global poverty. Since the late 1980s, the absolute number of people living on less than a dollar a day in Africa has grown five times more than the figure for Latin America, and twice that for South Asia. Social indicators have improved in many countries, but at a slower rate than in other regions. The structural adjustment policies promoted by the World Bank and International Monetary Fund (IMF) during the 1980s are often blamed for deepening poverty on the continent, particularly in southern Africa, although countries experienced reforms to different degrees.

The rolling-back of the state, in combination with trade liberalisation, has often reduced access to government services and support while at the same time increasing market competition. Small-scale farmers have seen the cost of agricultural inputs and consumer goods rising faster than the prices of agricultural produce, which has created a high risk, low return environment in which it is difficult for them to compete in either domestic or international markets. Scoones and his colleagues observe:

The gains from liberalisation in Africa have been patchy, limited or absent. Poorer farmers have lost the support once offered by (admittedly inefficient and often corrupt) parastatal marketing boards and government research and extension systems, but have rarely gained new support, markets or production opportunities. The consequence has been increased impoverishment for many, and growing inequalities between those who have gained and those who have been marginalised (2005:4). In many countries, farmers have ceased growing traditional export crops and commercial staple foods in rural areas remote from roads and urban markets, and both diversification into non-farm activities and migration of family members have become important livelihood strategies. Those living in urban areas have also seen increasing food prices, rising service charges, rising unemployment and declining gross and disposable incomes. Both the rural and urban poor have experienced cuts in public expenditure, especially in health, education and infrastructure, although spending per capita still tends to be higher in urban areas. In many cases governments' ability to provide services has been further constrained by very high debt repayments. Even after various rounds of debt reduction, sub-Saharan Africa still pays out more on debt service than it spends on health.

Economic hardship has had a profound affect on the linkages between urban residents and their kin and relatives in home areas. These linkages are traditionally very strong in most of sub-Saharan Africa, and represent a strategy to spread resources in a risk-prone environment. Research in Senegal and in Zimbabwe, however, shows that economic downturns have forced urban residents to decrease their financial support to rural-based relatives, with implications for the remittance economy on which so many of the rural poor depend. This trend is far from universal, however. In Botswana and South Africa urban dwellers' investment in home areas as a form of safety net, including assets such as livestock and housing, has continued despite growing uncertainty in the urban centres. Economic crisis may also be encouraging urban to rural movement, especially among retrenched formal sector workers, but the evidence for this remains anecdotal.

Many of these trends have been extended and compounded by the global balance of trade. The last three decades have seen economic stagnation in many African countries and a collapse in their share of world trade, which fell from around 6 per cent in 1980 to 2 per cent in 2002. This is largely due to African countries' continued reliance on primary commodities and a very narrow range of exports, which leaves them vulnerable to long-term declines and fluctuations in global prices. Africa's export prices are nearly four times more volatile than those of developed countries. Between 1980 and 2000, for instance, the price of commodities sugar fell by 77 per cent, cocoa by 71 per cent, coffee by 64 per cent and cotton by 47 per cent. Trade barriers, tariffs and agricultural subsidies have also played an important role. The heavy subsidisation of agriculture by many developed nations has also depressed world prices, making it difficult for African farmers to compete. This is compounded by import tariffs which further increase the costs to African producers. Internal barriers between African countries and Africa's failure to capitalise on intra-Africa trade also constrain growth. Such factors have in turn limited investment. As summarised neatly by the Tony Blaire's Africa Commission:

All this has had a knock-on effect. Investors, both domestic and foreign, see Africa as an undifferentiated whole – war in one country casts long shadows not just over neighbouring states but over the whole continent. As a result Africa seems to many outsiders an unattractive place in which to invest or keep their money. And what money is made in Africa is encouraged to flow out. Around 40 per cent of African savings are kept outside the continent, compared with just 6 per cent for East Asia and 3 per cent for South Asia. What is true of money is also true of people. Many educated Africans have over the years quit their homelands because they are frustrated at not being able to put their skills to good use. They can also earn more and have a better life elsewhere. Africa loses an average of 70,000 skilled personnel a year to developed countries in this brain drain. Zambia has lost all but 400 of its 1,600 doctors in recent years (2005:26).

Globalisation has at best done little to address these problems, and is at worst exacerbating them. The Economic Commission for Africa (2000) argues that although there are significant variations between countries and sub-regions, Africa's economy has been largely bypassed by the increase in global trade, investment and financial flows. Research in countries such as Kenya shows that poor Africans can benefit from globalisation (Box 1), but a recent study of rural Ethiopia suggest that such benefits may be confined to those areas with good infrastructure and communications. It suggests that already marginalised communities may often become more marginalised, as they are increasingly left out of the loop. There are also examples of the potentially negative effects resulting from international competition on the informal sector. In Kenya, for example, women who have traditionally produced sisal bags informally now face stiff competition in the form of cheaper imitations from south-east Asia, to whom they are losing market share, and as a result, suffering a loss of income (Carr and Chen, cited in Verick, no date).

Governance

Poor - and as suggested by the preceding sections, often constrained governance - also impact on livelihoods on the continent. Africa has for several decades been characterised by undemocratic governments, widespread corruption and ineffectual states. As noted by Africa Commission, many countries:

Have suffered from governments that have looted the resources of the state; that could not or would not deliver services to their people; that in many cases were predatory, corruptly extracting their countries' resources; that maintained control through violence and bribery; and that squandered or stole aid (2005:106). Whether by their own doing, or because of the pressures placed on them by the global community, many Africa governments have invested insufficiently in the fundamentals needed to promote and secure people's livelihoods, such as technology, health and education systems, roads, power grids, telecoms, affordable housing and other services. In many cases, this has been due to limited human and financial resources to implement programmes. These capacity constraints extend to policy planning, design and implementation. In their report, the Africa Commission (2005) argues that many countries have poor quality systems for the collection of data, without which government policies can neither be properly formulated nor accurately monitored. Civil servants, in both national and local government, also often lack the skills and training to analyse complex information or plan and budget effectively. This has been exacerbated by the 'brain-drain', which attracts those with the necessary skills away from the countries that need them. There are more African scientists and engineers, for example, working in the United States than in Africa.

Such issues not only have direct bearing on people's livelihoods, but also on countries' ability to prevent and respond to disasters. Africa's Regional Strategy for Disaster Risk Reduction, for instance, recognises that disaster risk reduction has yet to be institutionalised effectively, and that both technical institutions and risk management structures posses inadequate financial or human capacity to address risk or respond effectively. Inadequate financing is a particular problem, and arises from the low priority given to disaster reduction in national budgeting and a lack of dedicated disaster funding mechanisms. Where disasters occur, institutions are often overwhelmed, particularly given limited use of risk-spreading and transfer mechanisms such as micro-finance, formal insurance and private risk pools. Disaster risk reduction mechanisms also suffer from many of the same governance weaknesses plaguing development interventions more broadly, such as low compliance and enforcement of policies, laws, regulations, standards and codes.

Corruption is another important constraint. This particularly harms the poorest people, as it diverts funds away from service provision and may promote poor policies. The poor also have to pay a higher percentage of their income in bribes. In Kenya, for example, the rural population experience what Ade Freeman and his colleagues term a "multiple shaking down", where officials regularly require they pay a range of spurious fines and fees for services that should be free. A recent study by Kenya's Anti-Corruption Commission (KACC) shows that almost one third (31 per cent) of Kenyans seeking public services reported that they pay bribes, with Kenyans paying an average of US$ 30 in bribes annually. Poor people may be more likely to have to pay such bribes, as they are less able to navigate round the blockages and disabling contexts that everyone has to deal with while trying to make a living. Thus Freeman and colleagues (2004) argue that the potential for success in rural poverty reduction in Kenya will depend on the government's ability to address disabling and discouraging public sector institutional environments.

The HIV/AIDS Epidemic

High levels of disease, including HIV/AIDS, tuberculosis and malaria, also significantly undermine livelihoods on the content. The HIV/AIDS epidemic, in particular, has serious implications for livelihoods in Africa, especially in countries south of the Sahara.

While the latest statistics published by the Joint United Nations Programme on HIV/AIDS (UNAIDS) indicate worrying increases in infection in other regions, especially south-east Asia and the Russian Federation, sub-Saharan Africa remains home to two thirds (67 per cent) of those living with HIV/AIDS globally. Three fifths (75 per cent) of all those who died from AIDS in 2007 were also from sub-Saharan Africa.

HIV/AIDS directly undermines livelihoods in a range of ways, many of which are still poorly understood. Most fundamentally, AIDS-related illness and death drains household incomes and assets, and draws both income and labour away from livelihood pursuits. Financial strain, and the loss of productive adults diminishes social capital and assets, and strains cooperative livelihood strategies, such as labour pooling in the agricultural sector, as well as other kin and community safety-nets.

There have been very few studies on the impact of the epidemic on communities, but a handful of studies suggest that the implications of the epidemic may be community-wide in heavily affected regions. A study in Tanzania, for instance, showed that while almost one third (32 per cent) of households were directly affected by HIV/AIDS, a similar proportion (29 per cent) had been affected through ripple effects such as fostering of orphans, providing labour or cash to help care for sick individuals and assisting other members of affected households.

Less directly, the illness and death of educators, civil servants and leaders adds to the resource constraints that already exist, and is likely to increasingly strain the reach and responsiveness of government institutions. It may also impact on overall economic growth and performance, although the magnitude of the epidemic's effects on national and regional growth has yet to be proven.

ENVIRONMENT DEGRADATION, DESERTIFICATION AND CLIMATE CHANGE

Environmental degradation and desertification are already significant problems in much of Africa. It is estimated that two-thirds of African land is already degraded to some degree and that land degradation affects at least 485 million people, or 65 per cent of the African population. As much as two-thirds of Africa is classified as a desert or dryland and desertification appears to be increasing, particularly around the Sahara. Nigeria, for example, is reportedly losing 2 168 km2 of rangeland and cropland to desertification each year, while Ghana is thought to be losing roughly 20 000 hectares. Degradation and desertification are driven both by climatic factors that reduce rainfall – drought - and human factors, such as population growth, land pressures and poor land-use practices, as well as economic, technological, policy and institutional factors. Poverty is a key driver of both degradation and desertification, particularly where it encourages unsustainable farming practices. As noted in a recent Economic Commission for Africa report:

Without alternatives poor people are forced to exploit land resources including fragile lands, for survival (food production, medicine, fuel, fodder, building materials and household items). Given that most drylands in Africa are poverty hotspots as well, the risk of desertification is high in many of these areas, as the poor inevitably become both the victims and willing agents of environmental damage and desertification (2008:5).

Environmental degradation and desertification undermine livelihoods in numerous ways. Most directly, they decrease the productivity of the land, increase food insecurity and reduce the water available to humans, crops, livestock and industry. They also reduce biodiversity and result in the loss of ecosystem services that are vital to millions of poor. Overgrazing on rangelands, for example, not only reduces protective soil cover and increases erosion, it also causes long-term changes in the composition of vegetation that may encourage the proliferation of unpalatable plant species and reduce biomass productivity.

Climate change is also a growing concern. The weather is already becoming increasingly volatile in Africa, and it is expected that climate variability and the frequency and intensity of severe weather events will increase. Current predictions suggest a future warming across Africa of 0.2-0.5°C per decade. It is likely to get drier in northern and southern latitudes and wetter in the tropics, with significant variation within regions and countries. In many parts of the continent, lower rainfall and higher temperatures are likely to increase both aridity and the frequency and length of droughts. The sub-Saharan drylands are predicted to be amongst the most vulnerable to climate change. Studies in Sudan and Nigeria indicate that large, poor, growing populations living in dry climates that are highly dependent on farming and grazing for livelihoods, live on marginal or degraded lands and are subject to weak governance institutions will be most vulnerable to the changes in water balances brought about by climate change. Rising sea-levels, coastal erosion, saltwater intrusion and flooding will also impact on coastal communities and economies. In Tanzania, for example, it is estimated that a sea level rise of only half a metre would inundate over 2 000 square kilometres of land. Such changes pose additional threats to agricultural productivity, to food security, water and energy security and to health.

Much of the continent's vulnerability to climate change stems from its energy poverty. It is estimated that 89 per cent of the region's population relies on biomass—wood, animal dung or crop waste—and other natural resources for their main energy requirements. This dependency increases the vulnerability of much of Africa's rural population as traditional biomass scarcity is tied to the extremes of climate variability. The combination of climate-related threats to existing energy systems, coupled with the low efficiency of existing energy technologies and reliance on expensive, decentralised energy supply systems, further compounds the problem of ensuring local energy access and meeting demand – both of which are essential to the livelihoods of the poor.

Conflict

Conflict is another key driver of risk in Africa, and East and Central Africa in particular. As shown in Figure, analysis by the Food and Agriculture Organisation (FAO) shows that current or previous conflict is strongly associated with food insecurity and crisis. Similar analysis on the causes of famines in Africa during the 1990's, for instance, shows that only one of the eight famines experienced during the 1990's were mainly caused by drought, with the rest triggered by civil wars.

While in all other regions of the world have seen declining numbers of civil wars over the past thirty years, the number of conflicts in Africa has risen. In recent decades Africa has experienced more brutal coups, drawn-out civil wars and instability than any other part of the world. Some, like the violence in Darfur, has been high profile, but there are countless smaller conflicts, such as those between herders and cultivators which are equally vicious. More than one out of every three of the estimated number of people killed by conflict between 1994 and 2003 lived in Africa, and millions of lives have been lost. More people have been forced to flee their homes in Africa than anywhere else in the world. Two out of every five (43 per cent) refugees and internally displaced persons (IDPs) globally are in Africa, many of whom live in crowded and insecure refugee camps or already-overcrowded cities and towns. The 2005 Millennium Development Goals (MDG) Report identifies conflict as a major challenge to achieving the MDGs and establishes a clear link between conflict, poverty and disasters: Efforts to eradicate poverty and hunger are frequently set back by conflict and natural disasters. Hunger and poverty, in turn, can provide fertile ground for conflict, especially when combined with factors such as inequality, and make being prepared to cope with disasters more difficult.

Several features of conflict serve to undermine livelihoods and increase both the potential for disasters and vulnerability to them. Most fundamentally, injuries and deaths destroy one of the few assets available to many poor people in Africa, their labour, and as in the case of HIV/AIDS, deplete financial and social capital, impoverishing families reducing their capacity to resist and recover from disasters. Conflict destroys market, agricultural and health systems, as well as the economic and governance systems that provide employment, credit bases, insurance schemes and markets for produce and household items. It destroys essential and non-essential services, including early warning capacities and disaster response. It also displaces people from their homes and communities, while the deliberate destruction of society fragments communities and social safety nets.

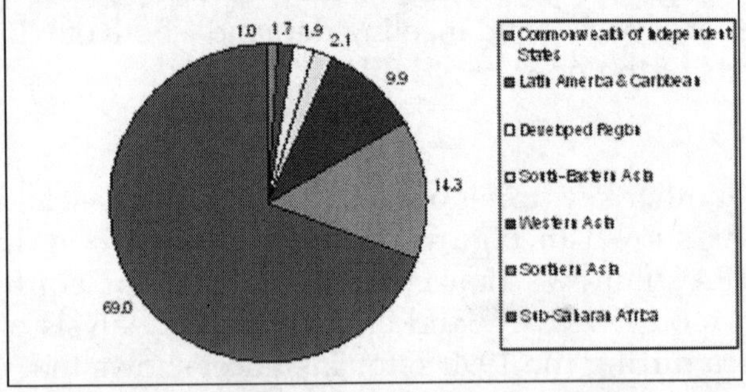

Fig. The proportion of estimated number of deaths in conflicts, 1994–2003 by region

Both disasters and stressed livelihood practices can also create and fuel conflict. Using data generated by the Conflict Data Project at the University of Uppsala, for instance, Robert Cincotta and his colleagues (2003) show that countries with limited per capita availability of cropland and/or renewable fresh water were one and a half times more likely to experience civil conflict between 1970 and 2000. While disputes over access to resources are rarely, if ever, the simple product of scarcity alone, they argue that they frequently collide with other features of economic, ecological and political environments to generate tensions – and are often emblematic of such tensions. Drought, environmental degradation and desertification, for instance, have fuelled conflicts in Central and East Africa. Land-related threats to traditional rural livelihoods, such as disputes over farmland distribution or the settlement of outsiders in traditional ethnic homelands also feature prominently in the evolution of recent civil conflicts. The conflicts in Rwanda, Burundi, the Democratic Republic of the Congo, Ethiopia, Sudan and Somalia, for example, have all had a resource component.

Rural Livelihoods and Risk in Southern Africa

In July 2002, the World Food Programme (WFP) estimated that 12.8 million people were in need of food aid in six southern African countries: Zimbabwe, Malawi, Zambia, Lesotho, Swaziland and Mozambique. The Disasters Emergency Committee (DEC) - made up of thirteen British charities, including ActionAid, Save the Children, Oxfam and World Vision - appeal calculated that an additional 1.6 million people needed WFP food assistance in Angola, bringing the total number in need of assistance to over 14 million people. At the time, the WFP described the shortages as "the worst food crisis in a decade"; the shortages were most severe in Zimbabwe, where almost half of the population were believed to require food aid, followed by Malawi where just over one quarter of the population needed assistance. In some countries, shortages persisted well beyond 2002/2003, with the WFP providing food aid to nearly 11 per cent of Malawi's population in 2005, for instance.

Table. Food requirements in Southern Africa in 2002

Population in need of food aid	% in need food aid aid	of	Metric tons of cereal aid up to March 2003	Cereal food aid as a % of national requirement
Zimbabwe	6 075 000	46	705 000	33
Malawi	3 188 000	28	208 000	11
Zambia	2 329 000	21	174 000	10
Mozambique	515 000	3	62 000	2
Lesotho	445 800	20	50 000	14
Swaziland	231 000	21	12 000	7
Total	12 783 000	22	1 211 000	13

The food shortages were triggered primarily by droughts, which in some countries combined with flooding. In Malawi, for instance, localised flooding in 1998 and 1999 combined with drought during the 2001/2002 agricultural year to produce a combined shortfall of nearly 600,000 metric tonnes of maize. The factors driving risk varied between countries. However, high levels of poverty, limited livelihood options, particularly a reliance on poorly diversified rain-fed agriculture, and the cumulative effects of repeated poor harvests were key factors, which combined political instability, poor governance, structural inadequacies in the agricultural sector and high levels of HIV/AIDS to make countries particularly vulnerable.

Livelihoods, Assets and Risk

As in other parts of Africa, agriculture forms the mainstay of both national and household incomes in southern Africa, increasing the magnitude of the hazards. Over 85 per cent of Malawians for, instance, rely on either commercial or subsistence agriculture for part or all of their income, while almost two fifths (37 per cent) of the country's GDP comes from agriculture. This dependence is compounded by under-developed service and manufacturing sectors, which limit off-farm work opportunities. Low levels of agricultural diversity and reliance on rain-fed agriculture also increases vulnerability to variations in rainfall, while skewed land-distribution mean that the most fertile land is held by a small number of wealthy farmers, with the majority of subsistence and small-scale farmers confined to less productive, marginal land, that is prone to degradation and climatic variations.

These factors, however, have been a feature of rural landscape for several decades, and while certainly part of several cascading vulnerabilities, do not appear to have been instrumental in creating the shortages. While the shortages of 2002 saw some of the highest levels of hunger in almost 50 years in Malawi, for instance, the proportionate shortfall in the harvest was significantly lower than during the major drought of 1991/1992, when no famine was reported (Office of the UN Resident Coordinator, cited in Garbus, 2003). ActionAid (2002) Malawi argues that one factor differentiating the 2001/2002 drought was that it followed a string of previous droughts that depleted physical and financial assets and left poor households with few resources to draw on. Others point also to growing population densities, have progressively allowed small-scale and subsistence farmers access to increasingly smaller and less efficient holdings. The HIV/AIDS epidemic has also significantly increased vulnerability to climatic variations. While all famines have long-term roots in uneven development, several commentators believe that the fundamental difference in the 2002/3 crisis was high levels of HIV/AIDS mortality, which both worsened and was exacerbated by the food crisis. James Morris, the Special Envoy of the Secretary-General for Humanitarian Needs in southern Africa, and Stephen Lewis, the Special Envoy of the Secretary-General for HIV/AIDS in Africa, concluded in January 2003 that:

HIV/AIDS is the most fundamental underlying cause of the southern African crisis. Combined with food shortages and chronic poverty, HIV/AIDS becomes even more deadly. The link between food security and HIV/AIDS must be fully recognised in all government, United Nations, international and NGO efforts to address food emergencies and in their support of HIV/AIDS-affected populations (Drimie and Mini, 2003:13)

Commercial and subsistence agriculture are particularly vulnerable to the effects of HIV/AIDS. There is debate as to whether the consequences of the epidemic are sufficiently far-reaching to fuel a 'new variant famine' or simply represent compounding shocks to fragile livelihoods, but most agree that AIDS depletes human resources, causes incomes losses and divert capital from agriculture, as well as having a range of other psycho-social impacts that affect productivity (Mutangadura, Jackson and Mukurazita, cited in

Drimie and Mini, 2003). Vulnerability assessments carried out during the 2001/2002 crisis show, for example, that income levels in AIDS-affected households were substantially lower than in other households in the same community. Human, labour and financial losses often result in a shift from cash crops to subsistence farming, and from labour-intensive to less intensive (and often less profitable) crops. These and other dynamics exacerbate food shortages by reducing:

- Food availability (through falling production, loss of family labour, land and other resources, loss of livestock assets and implements);
- Food access (through declining income for food purchases); and
- The stability and quality of food supplies (through shifts to less labour-intensive production) (Loewenson and Whiteside, cited in Drimie and Mini, 2003:19).

National Mismanagement and Governance

Several commentators also point to the role of mismanagement and poor governance in creating and magnifying the food shortages. In addition to factors like the relative neglect at the policy level of the smallholder agriculture sector, inaccurate crop forecasting, and a general failure to react to signals of an impending food crisis, ActionAid (2002) argues that the Malawi government's decision to sell off its 110 000 ton Strategic Grain Reserve in 2001 played a key role in translating drought into food shortages. Others pin-point the privatisation of the state maize marketing board, which resulted in the closing of the board's most rural, and therefore least profitable, depots. Prior to privatisation, the Board had had depots in many of the most remote rural areas, and provided a key safety-net for the rural poor in times of drought. As Owusu and Ng'ambi argue:

In the past, food shortages have been addressed through food aid from donors and government subsidies for basic food channelled through the grain board…In 1991-92, there was a severe food shortage in Malawi, with yields much lower than those preceding the current famine. However, the state marketing board, had depots in the most inaccessible rural communities and made food available at subsidized prices. This system has allowed the people of Malawi to survive the seasons of adverse weather, and the government corruption and mismanagement which has persisted through years of good harvest and bad. This year, however, no such safety net exists.

Craig Richardson (2007) makes the link between both the 2001/2002/2003 and ongoing food shortages in Zimbabwe to the government's radical land redistribution programme. He examines the impact of drought on Zimbabwe's GDP and highlights the changing vulnerability of Zimbabwe's agricultural sector to drought. Drawing on rainfall data for the period between 1960/1961 and 2002/2003, he argues that Zimbabwe has weathered significantly more severe droughts that those experienced in 2001/2002, with less impact to its economy, and from which it was able to rebound more rapidly. He attributes the change to the government's seizure of white-owned commercial farms and the associated neglect of agricultural inputs and infrastructure.

Most pertinently, he notes that while the seizures have led to "cascade failures" (2007:471) throughout Zimbabwe's agriculture-dominated economy, they have impacted most severely small-scale and subsistence farmers who used to obtain seeds, gasoline and other inputs from the neighbouring commercial farms. They have also displaced thousands of rural workers, many of whom have moved to urban areas in search of work, or adopted or expanded other livelihood options such as panning for gold in streams and rivers. In a prime example of the potentially negative feedback loops created by poor policy, however, he notes that such panning is speeding the silting-up of reservoirs used to irrigate remaining high-value crops such as flowers and vegetables, threatening too these livelihoods.

His conclusions are controversial. Jens Andersson (2007) argues that the association of food production with the commercial farming sector is overly simplistic. He argues that food insecurity in Zimbabwe is driven by two reinforcing trends: the expansion of maize production onto marginal lands and greater reliance on small-holder food production. Andersson maintains that the combined effects of the colonial administration's eviction of Africans into so-called 'communal lands' and ongoing population growth mean that the African population increasingly farm on poor, infertile, sandy soils prone to degradation, which leave their agricultural production vulnerable to adverse weather conditions. In parallel with these trends, the disruption of primarily commercial settler agriculture during Zimbabwe's liberation struggle, and its increasingly export-oriented focus, also led to reliance on weather-sensitive small-scale farming by the African majority for food – which has since the mid-80s produced nearly 60 percent of the country's staple food crop, maize.

He concurs, however, that changing government policies and Zimbabwe's declining economy have added to the vulnerability of food production to the climatic variations. He argues that the rapid growth in maize output the 1980s was achieved by an intensification of smallholder maize production through the use of hybrid maize varieties, mineral fertilizer, agricultural extension, agricultural credit systems and subsidised producer prices. This high-input farming regime was financed by a combination of agricultural credit, crop/livestock sales and investments from non-farm income sources, and made food security vulnerable to cuts in government support. In the late 1980s, agricultural credit for smallholders dried up, and, following Zimbabwe's adoption of structural adjustment policies in 1990, producer price support decreased, leaving smallholders increasingly dependent on their own non-farm income sources, such as remittances from primarily urban-based wage labour. However, these sources too have declined due to rising unemployment levels since the 1990's, while foreign exchange shortages have made mineral fertilisers increasingly expensive and scarce. Together, these trends have whittled away smallholders' production capacity and rural families' resilience against adverse weather conditions.

International Pressures

These examples highlight the role of higher-order political and economic processes in driving the food shortages. The structural adjustment programmes promoted by the International Monetary Fund (IMF) and the World Bank in the 1980's and the 1990's, for instance, have played a key role in increasing the vulnerability of the agricultural sector to variable rainfall. The gains from these programmes have been patchy, limited or absent. Such international players have also been directly involved in poor decision-making. The IMF, for instance, was instrumental in convincing Malawi to sell its grain reserves in order to balance its payments, even though there were already signs of food insecurity. Malawi's dependence on foreign aid has also left it highly indebted and subject to the "ebb and flow of resources as well as international donor trends", both of which hamper policy making and the ability of governments to respond to hazards.

CONFLICT AND RISK IN EAST AFRICA

Sudan has been plagued by resource-related conflicts for centuries and has been in various states of conflict for much of the last 50 years. The conflict in Sudan's eastern Darfur has been particularly severe and remains ongoing, despite the signing of the Darfur Peace Agreement (DPA) by one rebel faction and by the Government of Sudan (GOS) in May 2006. The conflict is broadly between the GOS and rebel opposition movements (the SLA and the JEM), and is being played out along ethnic lines between GOS-supported Arab militias—the Janjaweed—and other tribes that support the Sudan Liberation Movement (SLM). Attacks by both sides of the conflict have been associated with killings, burning of villages and looting of assets. The fighting has displaced an estimated 2.5 million people. It has killed thousands more, although the numbers vary widely from 98,000 to 400,000 or more (CNN, 2008). Many of these deaths have been the result of starvation and disease rather than fighting.

The Impact of Conflict on Livelihoods

Prior to the conflict, livelihoods in Darfur were based on a combination of farming, herding, trade and labour migration. The region was usually able to achieve food self-sufficiency in all but the worst drought years, with the more prosperous and surplus-producing parts serving as a source of both employment and grain for the poorer parts. However, population displacement, destruction of property, looting and restricted movement have devastated livelihoods, creating food insecurity and poverty and reducing households' resilience to droughts and other climatic conditions.

Mobile pastoralism, for instance, developed as a response to the region's low and erratic rainfall, the low biological productivity of the soil and cyclical droughts. However, the conflict – and other factors, such as a lack of policies and investment in services and infrastructure in rural areas - has blocked and restricted migration routes. This has cut herds off from essential water and pasture, while the concentration of people and livestock in safer areas has put unsustainable pressure on both services as well as natural resources. As the combined effects of the conflict and the negative policy environment have grown, "choices have diminished and today the population is in a state of sustained social and economic crisis".

The disruption of remittance flows from migrant workers has particularly increased the negative effects of droughts and other climatic variations. Migration and trade have for many years been central features of livelihoods in Darfur, providing up to 40 per cent of food and income. While millet farming was the dominant livelihood strategy, men tended to migrate to central Sudan or (less commonly) to the Gulf States in search of semi-permanent work, while women tended to migrate to South and West Darfur in search of seasonal agricultural employment between the months of November and January. Such migration was not subject to the vagaries of the weather, and in the past was an option even in periods of profound drought and famine. This allowed Darfurians to diversify their income sources and thereby increase their resilience to periods of hardship and food insecurity. The conflict since 2003 has it made it difficult for either people to leave or send back remittances, cutting off this vital source of support.

Patterns of aid to the conflict-zone have also affected livelihood strategies. While food aid to the region provides not only food for consumption and income, particularly for IDPs who have limited livelihood alternatives, it may in some cases actually undermined livelihoods. Sara Pantuliano argues, for instance, that the concentration of the WFP's food distribution processes around main villages and small towns often overlooked the nomadic population. Where aid reached such groups, this approach encouraged nomads to settle, increasing pressure on the local ecology and making it harder for communities to either farm or raise livestock. The WFP's approach also did little to support pastoralists' livelihood strategies. Drawing parallels with the WFP's experience during the famine in the early 80's, she notes:

When not overlooked, Beja pastoralists often had to accept distribution modalities that diverted them from their attempts to reactivate their traditional livelihoods system. Food for Work (FFW) schemes run directly by WFP in the southern part of the then Red Sea Province were almost all focused on the construction of social services (schools, health centres, latrines, hafirs or water catchment areas, etc.) or on agricultural activities. None of the projects bore any relation to the pastoral livelihoods of the Beja and they were seen as 'seriously detracting from the Beja's ability to rebuild their herds' (2007:S79).

The failure to follow the food relief provided in the 80s with appropriate rehabilitation and development strategies may also have impacted negatively people's livelihoods and reduced their resilience to both the current conflict and weather – a situation that is being replicated under current relief efforts. Pantuliano (2007) argues that food aid would have been more effective had it been part of a coordinated development process focused on a set of multi-sectoral interventions to address the root causes of livelihoods vulnerability, and that it should have been gradually replaced by appropriate development interventions. This failure is partly explained by the fact that agency personnel were not trained to make the transition from relief to recovery efforts. More instrumentally, she argues that political restrictions placed by donors on development funding for Sudan allowed only short-term emergency funding that did not allow the development of longer-term recovery and rehabilitation initiatives. The availability of food has in turn acted as a disincentive for government to develop more proactive measures to address food insecurity.

Both the effects of such weaknesses and the timeframes required to rebuild asset bases are illustrated by experiences of settled agro-pastoralists in Angara village in Nyala, eastern Sudan. The village was attacked, assets looted and the population displaced during localised tribal conflict between the Fur and Arabs in the late 1980s. Most did not return until 1994. By 2003, when the current conflict began, locals estimated that they had restocked to just half the pre-conflict levels of livestock, by selling cash crops and reinvesting in livestock. This suggests that recovery plans must contain realistic timeframes, despite the familiar pressures on aid agencies to implement and withdraw quickly from recovery projects.

Research in Kenya suggests that pastoralists take much longer to recover from asset depletion than crop farmers, as they depend on reproductive capital, which once eliminated takes a long period of time to recover. In this case, the unusually short drought cycles experienced in recent years – possibly due to climate change - mean that affected pastoral groups do not have time to adequately recover before another drought, suggesting the need for interventions that focus on preservation of livestock rather replacing lost stock.

The conflict has also had more complex, less tangible effects on local livelihoods and the potential for recovery in the region. Margie Buchanan-Smith and Sue Jaspars (2007) argue that the fighting has led to widespread pessimism; people have stopped making plans for the future or the recovery of their livelihoods, and are simply focused on how to survive under the current conditions. Their research amongst people affected by the conflict also indicates an acute awareness of how difficult it will be to coexist in the future with those who may have been on the other side of the conflict, or who may have exploited one, and how much needs to happen to rebuild the relationships necessary to revive livelihood strategies. Not everyone is affected equally, however. Sue Lautze (1997) notes that in any complex emergency there are both winners and losers; while conflict are associated with extreme impoverishment of vulnerable groups, they are also marked by massive accumulation by those with power. In GOS controlled areas, for instance, Arab pastoralists have benefited by demanding protection money from and the looting of non-Arab groups. Similarly, while most local markets have closed as a result of the conflict, many secondary markets have prospered due to declining competition, particularly in SLA-held areas, where secondary markets have often become important trading centres. These winning and loosing positions are not static, and often shift during the course of the conflict as areas change hands or circumstances otherwise change.

Research by Luka Biong Deng (2008) also suggests that poor households are not necessarily the most affected by conflict. His research in three communities in southern Sudan shows that while poor households are worst affected by shocks like drought, it is non-poor households who are often most affected by conflict. He shows that counter-insurgency militia drawn from the study communities – Dinka militias – were most likely to target better-off households, particularly those with livestock, the basis of the Dinka economy. Arab militia containing soldiers from outside the study communities, however, tended to target communities rather than individuals, in which case the effects were spread more evenly – although non-poor households still tended to fare slightly worse. In many cases, the theft of livestock pushed non-poor households into poverty. Deng argues that this "curse of assets" (2008:386) highlights the need for more nuanced analysis of the relationship between risk events, livelihood strategies and vulnerability, and underscores the importance of understanding not just risk, but the nature of risk and its effects on different groups.

The Impact of Livelihoods Strategies and Disasters on Conflict

The pursuit of some livelihood options in the face or resources scarcity can also generate conflict at various scales. In Sudan, the collapse of the pastoralist economy in some areas has encouraged pastoralists to adopt other livelihood strategies, such as collecting and selling firewood, which has brought them into conflict with other groups, like IDPs who have adopted the same strategies. In some places, overcrowding and competition for work and basic services since the influx of IDPs have also created tensions between them and the resident population.

Sudan is not unique in this respect. In Kenya, for instance, competition over access to watercourses and forage for livestock between the pastoralist Maasai and sedentary farmers has resulted in the several recorded conflicts between pastoralists and farmers, and among neighbouring pastoral communities. Conflicts are particularly common during droughts, when competition over grazing and water resources intensifies, and has in certain cases involved cross-border fighting. Insecurity, and the perception that areas are inadequately policed by the authorities, encourages communities to arm themselves, increasing the intensity of potential conflicts.

These conflicts have been influenced by changes in national land tenure policies, which have favoured the privatisation of land and encouraged the fragmentation of the communal holdings used by the Maasai to graze their cattle. The fear of loosing their land rights, and the potential profitability of farming around wetlands and watercourses, has encouraged many landowners to sell and subdivide their land. The amount of land under irrigation in one district, for instance, expanded from 245 to 4 768 hectares between 1973 and 2000 (Maitima and Olson, cited in Orindi et al, 2007). This has directly reduced access to streams and wetlands, dry season grazing areas and migration corridors for cattle, and has not only increased competition but also meant that pastoralists must move their herds further in search of water and fodder, contributing to the deterioration in the productive capacity of their livestock, particularly in dry years.

The Need for a Long-term View

The two-way relationship between conflict and livelihoods suggests the need for interventions that are mindful of the big picture. Governments and those providing aid should work towards strengthening local livelihoods in ways that not only support food security, but also economic, community and political security. Funding also needs to support longer-term rehabilitation and development efforts that promote sustainable livelihoods, rather than purely reactive interventions.

In northern Uganda, for example, the International Rescue Committee (IRC) has shifted to a more long-term approach focused on small-scale interventions that develop market linkages, with a view to supporting the regeneration of the economy and motivating individuals to become more productive. Its interventions aim to improve access to credit, the ability to manage earned income and develop tradable or marketable skills, as well as the demand for such skills or products. It also seeks to promote more supportive government regulations. This strategy has been implemented through Village Savings and Loan Associations (VSLAs), opening up land for agricultural production in cooperation with the military, the government and local landowners, providing business training and linking farmers with guaranteed markets in the private sector. All of these activities are targeted at the community at-large, are demand-driven and propose alternatives to traditional emergency-style economic support.

Rachel Locke (2006) argues that an emphasis on demand-driven skills is essential. She notes that many humanitarian interventions provide both short and longer-term support without consideration of market demand or systemic constrictions on economic activity. She notes, for instance, that hundreds of formerly abducted children have received training in tailoring, carpentry and brick-making, but without adequate attention to the implications of such interventions, this training has simply saturated the market for these skills, reducing economic opportunities for both these new and existing workers.

URBAN RISK IN SOUTH AFRICA'S WESTERN CAPE PROVINCE

Urban areas in the Western Cape are the site of ongoing, chronic hazards, from poor sanitation, to crime and violence and diseases such as HIV/AIDS that do not fit the traditional definition of 'disasters' but that fundamentally undermine the well-being and livelihoods of the poor on a daily basis. It has also experienced several conventional disasters that have exceeded the capacity of local authorities to respond. Between March 2003 and April 2005, for instance, three severe winter storms triggered widespread flooding and mudslides in various parts of the province. These resulted in very few deaths - ten altogether – but created combined direct losses of well over R 364 million (US$ 47 million). In both 2003 and 2006, the extent of the damage and displacement exceeded local response and recovery capacity, and several of the worst affected areas were officially declared disaster areas. While the storms affected both rural and urban areas, the effects were felt most keenly in urban areas, where poorly planned and managed urban growth has left poor households extremely vulnerable to the effects heavy rainfall.

Livelihoods, Development and Migration

Very high levels of migration have played a key role in driving urban risk in the province, which is in turn intimately bound up with people's efforts to secure livelihoods for them and their families. There have historically been fewer economic and other opportunities available to those living in rural areas, and oscillating migration has been a way of life for South Africa's rural poor for decades. Historically better resourced than most other provinces, and with some of the highest economic growth rates in the country, the Western Cape is a popular destination for migrants from surrounding provinces, and to a much lesser extent, elsewhere in South Africa and the continent. Between 1985 and 1990, approximately 30 000 people moved in the province each year. This rose to as many as 41 000 people between 1991 and 2001. In-migration has slowed in the last seven years, but it is expected that the Cape will gain 177 000 new residents by 2015, at an average of 11 800 people per annum from 2001 (Western Cape Provincial Treasury, 2006).

The majority of those migrating into the province are poor African families and individuals from the Eastern Cape, attracted by the jobs offered by the buoyant economy and substantially higher income, education, health and social development levels in the province. Access to infrastructure is becoming an increasingly important pull factor, particularly as urban labour markets have struggled to absorb the large numbers of migrant work-seekers. As one commentator observes:

South African urban and rural unemployment have reached levels where there may be little advantage for the rural unemployed in moving to the city to look for work. To an important extent, public goods that provide basic needs such as water and energy have become a second best substitute goal. Migration to advantaged areas — or at a minimum to centres that offer access to services and mass transport — may not provide an optimum income, but it allows the household to cut its transport costs and its labour time loss so as to put more resources into cash earning.

Even though many migrants never find jobs to support their households, informal business and other non-wage livelihoods options are more sustainable in these areas than in isolated and marginalised ones with high unemployment and resulting low cash flow.

However, Catherine Cross (2001) argues that many analyses of people's reasons for migrating are overly simplistic. While it is assumed that the rural poor want to migrate to the cities, research suggests that many would remain in rural areas if there were adequate resources available to them there. Surveys by the Development Bank of South Africa, for instance, show that three out of four (76 per cent) rural respondents wanted to remain in their area, despite high levels of poverty. That so many do move underlines the point that the poor migrate in terms of how they see their best opportunities and what information and contacts they have available – and that they migrate to the cities only when they see a clear advantage and have the means to do so. This suggests that improving access to services and opportunities in rural areas would improve livelihood opportunities and reduce levels of population movement.

Urbanisation, Service Delivery and Risk

The inflow of migrants has led to rapid expansion of urban settlements. George, one of the Western Cape's largest secondary towns, provides a good example. As shown in Figure, the extent of urban development in George and its surrounds nearly trebled over the 47 year period between 1957 and 2004 – with the bulk of this expansion having occurred since 1985, when the Apartheid government began dismantling restrictions on the movement of the so-called 'Black' population. While the overall urban area of George doubled between 1957 and 1985, from 7.39 km2 to 15 km2, it expanded by 600 per cent between 1985 and 2004, reaching 90 km2 by 2004. This represents an average annual increase of 3.9 km2 between 1985 and 2004 – fifteen times higher than the 0.3 km2 urban growth rate for each year during the previous period.

This rapid growth has resulted in a measurable increase in hard surfaces, and with it the risk of severe rainwater run-off that now exceeds both the technical and infrastructural capacity of existing services and infrastructure to manage. This is exacerbated by local government's failure to match growth with strategic investments in the redesign and maintenance of critical infrastructure, as evidenced by repeated losses to roads and storm-water systems, as well as sewage treatment and water supply plants. The increase in population has also resulted in urban sprawl into ecologically and geologically fragile areas, destroying protective ground covering vegetation and increasing the risk of landslides.

Urban planning has also failed to keep pace with population growth. While the South African government has committed itself to ensuring that all South Africans have access to descent housing, it has been unable to cope with rapidly growing - and often unplanned - demand. Between April 1994 and April 2006, the national Department of Housing built 20 747 low-cost housing units in Eden Municipal District, where George is located, but there remains a considerable housing backlog, with another 17 169 – 43 per cent of the total planned units – still outstanding at the end of 2006. As elsewhere in the country, this has led to the growth of informal settlements on the urban fringe, and to a lesser extent, in and around the city's wealthier suburbs. Unlike formal housing areas, which originate through planned urban development, informal settlements are entirely unplanned and have little or no infrastructure. Dwellings are built of cheap and available materials, such as iron sheeting, wood and plastic and are only nominally, if at all, weatherproofed.

Service delivery too is inadequate. While the Western Cape performs better than other provinces on most service delivery indicators, many communities have limited access to basic services. The Department of Local Government (DPLG) estimated in March 2005 that only four out of five poor households were receiving water (84 per cent) or electricity (84 per cent). In some municipalities the electricity backlog was as high as 24 per cent. In George, the DPLG calculates that, as of the last census, only four out of five (74 per cent) households had access to piped water on their property, were connected to sewerage system and had access to telephone and electricity services, while only one in five (86 per cent) had access to solid waste removal services. Inadequate service delivery increases both people's exposure and vulnerability to hazards. Poor drainage infrastructure, for instance, prevents flood waters from draining away, while rotting human and household waste clogs drainage ditches and contaminates standing water with germs and poisonous chemicals that cause disease.

These delivery failures are most fundamentally the product of limited financial and human capacity. Local municipalities have all undertaken to provide a free basic water supply and electricity to impoverished households, but simply do not have the necessary resources to meet demand. At a deeper level, they also reflect inadequate cognisance of both disaster risk and disaster risk reduction in urban planning and service delivery. South Africa's Disaster Management Act of 2002 explicitly calls on governmental roleplayers to reduce the vulnerability of disaster-prone areas, communities and households, and mainstream risk reduction into development planning. However, the ongoing vulnerability of communities and the repeated losses incurred in the province suggest that this is not happening. It also evident there is nominal, if any, interface between urban planning and risk reduction processes, with new developments frequently designed and constructed in ways that not only fail to reduce risk but drive it.

Urban Livelihoods and Risk

Several features of the urban poor's livelihoods make them vulnerable to the effects of storms. Urban households are often operating at the limits of their capacity, as they attempt to juggle multiple stressors such as unreliable incomes, food insecurity, diseases born of overcrowding and poor service delivery, and the effects of crime and violence (DiMP, forthcoming). They also tend to be more dependent on the cash economy than their rural counterparts, and despite often high levels of unemployment, experience more financial demands such as rents and the costs of food. Migration also often results in the loss of assets, as migrating households leave behind any land and grazing entitlements and lose social capital in the form of networks and social connections that formerly backstopped shortfalls in household income, making them highly vulnerable to economic shocks. Urban 'communities' are also often low in cohesion, turbulent and unstable, and subject to lower levels of social control, making it more difficult to re-establish such networks.

Hazards such as flooding erode whatever attempts have been made to accumulate resources and savings. The 2006, for instance, not only destroyed dwellings, but also damaged bedding, clothing, furniture, food, identification documents and other valuables, that often needed to be replaced. Many of those affected by the flooding also had to miss school and work, either because it was not safe to travel there, or because they had to rebuild or repair their dwelling.

An impact assessment by the Disaster Mitigation for Sustainable Livelihoods Programme (DiMP) at the University of Cape Town found that people tended to loose two to three days' work, and anything between one third and one half of their usual income. Those running small enterprises, or storing goods in home, also lost sources of livelihood. Small-scale farmers in one settlement alone lost livestock to the value of R 39 500 (US$ 5000). Local clinics also recorded higher than average numbers of children presenting with diarrhoea, which is often associated with contaminated water. This highlights not only to the health implications of the flooding, but also how the existing 'everyday' hazards born of poor service delivery, such as poor sanitation and water supply, can compound the effects of larger, less frequent events.

There is frequently a gender component to risk and vulnerability. Women-headed households were often worst affected. Interviews and field observations suggest that older women over the age of 50, particularly those with young children or grandchildren, were the most vulnerable. These households reported very low household incomes, and could only afford the cheapest building materials and building sites (Box 3). The number of such 'granny-headed' households is rising throughout South Africa, largely as a result of the HIV/AIDS epidemic, suggesting gender and age will become in increasing important factors in assessing and responding to risk in South Africa.

Social Insurance and the Accumulation of Risk

Very few low-income households can afford private insurance premiums in South Africa, and the poorest and most vulnerable households almost always bear the greatest loss burden following a disaster. Some of these costs are absorbed by governmental and non-governmental relief organisations, which provide emergency shelter, food and blankets, and in some cases, rudimentary materials to help people repair and rebuild their homes, but such assistance is generally relatively superficial. Moreover, while there are often small-scale rotational credit and savings schemes - locally referred to stokvels – operating in poor communities, it is unclear to what extent such schemes serve as a safety-net and buffer where whole communities are affected by severe weather.

DiMP's research showed that the costs of flooding often exceed households' average monthly income and that would take some time for households to save enough money to cover the extra expenses. While those living in informal dwellings reported an average income of approximately R 1 000, for instance, their average reported expenditure amounted to almost R 900 (US$ 118), leaving little spare money to spend on replacing or repairing damaged property. This is especially so when one considers that many of the same households have experienced repeated losses associated the multiple extreme weather events in the province since 2003. For these households, losses of even a few hundred rand can progressively increase their vulnerability to the effects of flooding and other hazards over time.

Rather than helping households to cope and building resilience, responses by government and other agencies may compound these processes. In 2006, the lack of a standardised, systematic and rigorous household needs assessment protocol to determine which communities required relief, such as food, blankets, toiletries and temporary accommodation, resulted in many of the hardest-hit households not receiving even the most basic assistance. Despite the apparent impact of the storms on poor communities, the relevant government department determined that "there was no need" for social relief during and following the floods. In justifying this decision, they argued that the public and business communities had already contributed sufficiently to meet the needs of the affected, making relief unnecessary. It is particularly incongruous that poor households did not receive social relief, when municipal and provincial departments sought emergency funding assistance from the National Treasury, while wealthier residents and businesses sought insured assistance to cope with and recover from the floods.

4

Drought Preventability and Prediction in a Changing Climate

INTRODUCTION AND MOTIVATION

Drought is a phenomenon that is basically linked with a sustained lack of precipitation, and in some case also with excess evapotranspiration. It is difficult, however, to precisely define drought, since its critical features are at least somewhat dependent upon the features of the region in which it occurs, and the sector of society that is affected.

The occurrence of drought is nonetheless a ubiquitous feature of the global water cycle. Drought does not necessarily lead to an overall change in the magnitude of the global water cycle but it of course affects the regional cycling of water. Droughts are recurring aspects of weather and climate extremes as are floods and tornadoes, but they differ substantially since they have longer durations and lack easily identified onsets and terminations.

Drought can be considered as an anomaly within the atmospheric, surface and sub-surface cycling of water and energy. This anomaly may be initiated through large to regional scale atmospheric processes, and is enhanced and maintained through regional to local atmospheric, surface hydrology, land surface and groundwater feedbacks that operate at various time scales over growing season and snow-covered periods. The atmospheric process may in turn be linked to long-lived SST anomalies, which can act to sustain droughts over many years.

Most regions of the world are affected by drought from time-to-time. Although it tends to be most common and severe over the central regions of continents, there are many exceptions to this. Drought has been and will continue to have some of the greatest impacts of climate. Drought has brought ancient civilizations to their knees and leads to famine, food scarcity, loss of lives and property. Evidence suggests that the area affected by drought has been increasing on a global basis. Furthermore, many climate predictions anticipate a further increase in drought-affected regions in the future. Impacts from drought can be reduced if there is forewarning. This can be on relatively short scales for weather conditions within a drought to decadal scale expectation involving natural decadal variability as well as climate change.

Numerous studies of drought have been and are being conducted. Many of these are concerned with its characterization ranging from estimates of its global distribution down to detailed studies of individual events. Other studies seek to better understand the large and small-scale factors that lead to, sustain and end drought. Other efforts strive to mitigate the impact of drought.

However, despite the enormous economic, environmental, and societal impacts of drought, there has never been a coordinated and integrated drought research program in the world. Given the importance of this extreme form of climate variability, it is critical that it be studied appropriately, with the hope to better understand its nature and anticipate its occurrence on short and long time scales.

An effort devoted to improved prediction of drought at various temporal scales is a natural undertaking for the World Climate Research Programme. WCRP has an existing network of researchers around the world who have the necessary expertise in the many aspects of drought and WCRP's mandate is to address such issues of great importance to society towards the development of a common framework and new drought-related climate services. This should serve as the basis for devising effective intersectorial communication and an improved drought early warning system for drought mitigation.

USER REQUIREMENTS FOR DROUGHT PREDICTION INFORMATION

The provision of relevant drought information to inform sector-specific decisions presents challenges that transcend scientific advances in the study of climate. From a user perspective, drought becomes manifest through its myriad impacts and those impacts vary markedly across sectors, timescales and location. It is for these reasons that drought eludes a universal definition. For example, a hydrometeorological definition of drought found useful to the agricultural sector in a given region may be deemed largely irrelevant to a water manager in the same region.

It is for these reasons that drought is generally classified into three major types (earlier reference? Keyantash and Dracup 2002): meteorological, agricultural and hydrological. Meteorological drought results from precipitation deficits, while agricultural drought is identified by total soil moisture deficits, and hydrological drought is related to a shortage of streamflow or runoff. All three types are ultimately generally related to a prolonged period of deficient precipitation (compared to the expected value at a given location), although evapotranspiration deficits can be critical for agricultural and hydrological droughts, and in some cases even more relevant than the precipitation deficits. Another important distinction between these drought types are their characteristic timescales, which can vary substantially. A single month of deficient rainfall can adversely affect rainfed crops while having virtually no impact on a large reservoir system. Yet, across classifications drought can have major societal impacts.

Drought was a major contributing factor to the famine in Ethiopia in the early 1980s where it is estimated more than one million people died. Drought results in billions of dollars in crop losses around the world each year and affects more people worldwide than any other natural hazard. Droughts can also lead to building damages and health impacts through an enhancement of temperature anomalies during heat waves (Seneviratne et. al. 2006a, Fischer et al. 2007). With increasing demands on water supplies the development and longevity of drought is a major concern in many areas around the globe. Given this context it is not surprising that a diverse set of user needs exists when it comes to providing timely drought-related information and predictions necessary to inform decision-making even before consideration of climate change.

This section emphasizes key considerations for the provision of meaningful information to those who manage drought-sensitive systems. That is, it addresses some important "user" requirements for drought information. These requirements are many and span time scales ranging from days to decades, and include new products and modeling tools that allow a seamless prediction approach (e.g., for seasonal-to-decadal forecasting of precipitation). To ultimately be effective, the development and provision of such information cannot proceed in a unidirectional manner; it requires an active, two-way engagement between those knowledgeable of climate science and the potential users of the knowledge gained by that science. Indeed, such a process is better described as the development of a decision support process, not just development of drought information. Identifying the multiple decisions and decision-makers in need of support is a challenge in its own right.

A thorough exposition on this subject is a major undertaking and beyond the scope of this paper. Instead, presented here is a set of criteria that serve to help frame the issues. These concepts largely originate from advances made in the development and practical application of seasonal climate forecasts in recent decades. They are extended here to include specific consideration of drought and the prospect of climate change. Four main themes emerge in this discussion: 1) the need to develop drought forecasts and information that are relevant to decision makers; 2) the need for a meaningful and ongoing engagement between providers of drought information and the users of that information; 3) the effective communication of drought information, and 4) demonstrable outcomes showing the utility (or at least potential utility) of using drought-related information in the decision making process. In fact, previous work has shown the relevance of these four concepts to both seasonal-to-interannual prediction and considerations of climate change.

THE NEED FOR RELEVANT DROUGHT INFORMATION

Relevant Predictands and Time Scales

From a user perspective, information relevant to drought management ultimately depends on what is being managed. In water resources, for example, it may be streamflow. In such a case predictions of that specific hydrologic variable are much more relevant than, say, predictions of seasonal precipitation, a hydrometeorological drought index, soil water balance, etc. In many cases, forecasts of the most relevant variables are not currently available. In such cases it is necessary to undertake research in order to quantify the relationship between the available predictor (e.g., seasonal precipitation, new meteorological "drought" indices that go beyond simple precipitation-based indices) and the predictand of interest (streamflow, crop condition, etc.) if the predictions (or even monitored conditions) are to be incorporated into decision-making. In some cases it is necessary to identify important threshold values rather than considering drought on a continuum. This could include consideration of multiple variables as well, such as drought condition, temperature, and humidity in the case of forest fire danger. And in general, the necessarily probabilistic nature of drought predictions must be clearly understood by users of that information.

The issue of time scale is another clear consideration. As drought is generally viewed as a slow onset phenomenon it may appear that the most relevant time scale for drought predictions is seasonal (or longer). However, particularly in the agriculture sector, shorter time scale meteorological information is also highly relevant. This ranges from short-term precipitation ("weather") forecasts covering just a few days to sub-seasonal outlooks of the soil water balance. In other cases, users are interested in predicted changes in the statistical characteristics of daily weather within a season (and under climate change), such as length or frequency of dry spells. If a crop model is being used to predict yields, it is often necessary to disaggregate seasonal forecast information (of precipitation and temperature) into changes in the statistical characteristics of daily weather as the latter are necessary inputs to the crop model (e.g., Hansen et al. 2006).

Hydrologic and land surface models also require inputs at these shorter time scales. On the seasonal time scale recent work by Robertson et al. (2008) has shown that in some regions the predictability of rainfall frequency is greater than the seasonal amount. This is of interest to practitioners of rainfed agriculture as is the potential prediction of the onset of the rainy season which according to recent work shows some promise for, at least in specific areas. The start of the planting season in many rainfed areas follows the onset of seasonal rains. In all cases, given the multiple constraints on decision-makers timely drought information and forecasts are essential if they are to allow for sufficient lead-time in order to take action.

Naturally, there are multiple demands for drought information which exist on time scales longer than seasonal-to-interannual. These include drought variations on decadal timescales and of course, the possible impacts of climate change on droughts. Based on historical observations and proxy climate data (such as dendrochronologies), there exists a large body of work documenting a diverse set of drought events that have occurred on decadal (or longer) time scales. Such information can be of great practical use to decision makers (water managers, the energy sector, infrastructure planners, etc.) provided it is expressed in terms relevant to the specific sector considered. For example, historical, observed flows in the Colorado River in the US have been successfully combined with tree-ring estimates of past precipitation variability to provide probabilistic estimates of the ranges of future flows in the Colorado during unusually dry or wet "decades".

Such statistical approaches can be extended to consider climate change by including projected trends in precipitation in the analysis (Rajagolpolan et al. 2009). In Australia, such an approach has been used to infer potential impact from rising temperature and rainfall changes under global warming on future inflows to Australia's longest river, the Murray-Darling River. Uncertainty in the projected trends can be assessed through an ensemble approach to provide the user with a range of future scenarios. These flow scenarios can then be used as input to a water management model to examine their effect given various, specified allocations (often for multi-purposes ranging from domestic consumption, to agriculture, to environmental protection, to recreation). Indeed, system sensitivity studies are one approach to provide decision makers with information on the potential effects of climate variability and change across multiple time scales. Such approaches need not include detailed (but uncertain) downscaled climate projections from climate models (e.g., Dessai et al. 2009). Rather, under a range of imposed climate scenarios decision makers can see how a (calibrated) model system responds. One can also assess how large a forcing is necessary to substantially alter the system can be assessed. Alternatively, given a capacity of a system (e.g., water supply), one may examine what the maximum severity and/or duration (of drought) are that will critically change the system, and then assess such probability. Naturally this approach is not limited to water resources. Examining how crop yields in a crop model respond to different sequences of dry spells or to an increase in rainfall intensity or in the length of (imposed) dry spells is one of many additional examples.

To be sure, system sensitivity analyses are not to be viewed as the single approach to convey to users the potential impacts of climate change. Climate model projections certainly have a role to play and a fundamental research challenge is to explore how uncertainties in climate projections can best be communicated to users. There are significant limitations and uncertainties ranging from spatial resolution, model physics, greenhouse gas emissions, and natural fluctuations. Firstly, for a regional application, the spatial resolution of climate model outputs is often far too low, hence the need to downscale. Secondly, we do not have precise knowledge of many climate processes and the necessary parameterizations of these processes are far from perfect. For example, model representations of cloud physics, various effects of anthropogenic aerosols, chemical ozone and its interactions with climate, carbon cycles, and land surface processes and their interactions with the atmosphere are areas in which significant uncertainty exists. A consequence is that given an identical climate change emission scenario, the response from one model to another varies vastly, and there is no objective way to place a higher confidence in a particular model. Thirdly, there is no certainty as to the level of various greenhouse gases in a particular future time, and much will be determined by the global mitigation actions. Finally, there is no obvious way to determine the phase of multi-decdal oscillations at a given time in the future.

It needs to be kept in mind that even in the realm of climate change, shorter time scales need to be considered. For example, quantified estimates of the changes in decadal, annual, or even seasonal precipitation from global climate models may have only limited utility for many users. As mentioned, in agriculture (or hydrology) it is changes in the sequences and/or intensity of higher frequency "weather' events that are often most important (and which "build" the temporally integrated changes on longer time scales). In addition, for many decisions such as the building of new infrastructure or urban planning, the most relevant time horizon is on the order of the next 2 to 3 decades. Projected changes in the behavior of drought at the end of the current century relative to the current climate are not necessarily relevant, and certainly not optimum. Indeed, there is a substantial, emerging demand by decision makers for decadal scale drought prediction, which targets only the next few decades. In such cases "natural" climate variability on decadal time scales may very well dominate, though, through nonlinear effect could be considerably exacerbated by any anthropogenic climate signal. Decision makers need to be made aware of both sets of information, where possible.

Relevant Spatial Scales

Most drought-sensitive decisions are made for specific geographic locations and thus actionable drought information needs to be made available at a relevant spatial resolution. Of course, in practice, the spatial resolution of information desired by decision makers may not match that which can be reliably provided. This applies to mapping of observed drought conditions as well as forecasts. Climate models used for seasonal to annual climate prediction as well as climate change projections are typically run at fairly coarse resolution (relative to the characteristic spatial scales of local variations in land cover, topography, microclimate, or the scale of individual farms, etc.).

This was one of the motivations for the development of spatial downscaling techniques for seasonal climate forecasts as well as climate change scenarios. Both statistical techniques and regional (dynamical) models have been used for these purposes. Naturally there are caveats to both techniques. First, in addition to their coarse resolution, the global climate models are not perfect representations of the climate system. They often contain systematic biases (in amplitude, or space) and non-systematic errors that, particularly when using a nested regional model, can influence downscaled results. Similarly, regional climate models also contain various errors and therefore require calibration with observed conditions before being used in applied settings. Since observations are generally sporadic in spatial coverage (especially for soil moisture, e.g. Seneviratne et al. 2010), even a calibrated regional model will still contain errors. When using statistical downscaling approaches it is essential not to overfit the model giving users an inflated sense of predictive skill. Statistical models need to be carefully cross-validated. Overfitting is in particular a problem in the context of climate change, since current climatic relationships may not apply to future conditions.

The upshot of downscaling is that high spatial resolution for drought does not necessarily equate to highly reliable information. And in some cases high spatial resolution may not be necessary. For example, there are several studies which document the successful use of global climate model information to directly predict streamflow (e.g., Brown et al. 2009; Landman et al. 2001). This approach avoids the "disaggregation, re-aggregation" problem of spatially downscaling a global climate model and then integrating the results over a watershed. The users of drought information need to know that such techniques are available, with demonstrated forecast skill. This is also the case because drought anomalies are often of large scale, e.g. when associated with circulation anomalies, such as during the European summer drought and heat wave. Thus high-resolution information is less critical for droughts than for some other weather and climate processes, such as heavy precipitation events.

APPOINTMENT WITH DECISION MAKERS

In addition to being relevant, drought information needs to be credible to users if they are to act on it. It needs to be transparent to users that objective, credible methodologies were employed when generating the information and providers need to clearly document the skill of their predictions if they expect them to be incorporated in decision making. However, making drought-informed decisions requires more than just the delivery of quality controlled climate information in a format understandable to users. It is essential that the decision makers themselves play a role in product development. For example, highly skillful and carefully cross-validated precipitation forecasts could be developed for a particular watershed but if a local water manager cannot incorporate that information into their decision-making process it has limited value at best. There needs to be a fundamental understanding on behalf of the drought information providers of the specific decision-making process being considered. In other words, to truly be effective, drought information and predictions (on any time scale) need to be transferred to users through a process of decision support, not just information supply (NRC 2009). This requires a continuous engagement between climate information providers and the potential users of their information. In the US, the Regional Integrated Sciences Assessments (RISAs) provide a means for such engagement. The envisioned National Climate Service in the US and other countries (as testified by the international declaration crafted at the WCC-3) will need to function in this decision-support capacity as well. Outside the US engagement between drought information providers and decision makers may at least be initiated at regional climate outlook forums (RCOFs). Again, sustained engagement is most likely to lead to useful drought-related product development.

EFFECTIVE COMMUNICATION OF DROUGHT INFORMATION

Currently, many weather and climate services around the globe provide some type of general information on "drought". This typically is in the form of various monitoring products such as accumulated precipitation over a given time period (season, past 90 days, past 12 months) relative to the long-term average and other aspects of hydrometeorological variables such as snow pack, surface water supply, and streamflow. Predictions (and climate projections) are most often made just for precipitation although soil moisture forecasts are becoming available for some regions and streamflow forecasts based on monitored snowpack have been made in many locations for some time. Most often such products, however, are designed for general use. To optimize its utility in a decision-support framework, various types of drought information ultimately need to be "tailored" to meet more specific needs. The challenge here is that the needs are many, and highly varied. Engagement with specific stakeholders is a necessary first step to providing more useful information as previously discussed. Specific challenges include 1) the need to synthesize all relevant drought information in an objective form, (current official US drought monitoring and outlook products are one example of this), and 2) an objective verification of drought products.

DEMONSTRABLE UTILITY OF DROUGHT INFORMATION

The ultimate uptake and use of specific drought-related information and forecast products by decision makers requires that they can be shown to have some tangible, documented benefit – or at least potential benefit. Such evidence may be provided in the form of a relevant case study for a particular sector and region, such as seasonal streamflow predictions and water management in Northeast Brazil or the northern Philippines. This has been an emerging area of inquiry over the past few decades in the realm of seasonal-to-interannual predictions with an increasing number of studies being published, sometimes with recommendations for approaches to the problem more generally. For the information provider and user alike it may also be insightful to consider relevant reviews of lessons learned from previous or ongoing efforts. Such studies must be relevant to specific decision makers, however, if they are to assist in the uptake of drought information. Alternatively, through engagement between scientists and decision makers, prototype products can be developed and subsequently used in a "hindcast" mode to show decision makers how such information could be used beneficially in their specific setting (and set of constraints). For example, an effective way to demonstrate the importance of the drought information is through use of historical examples when it is - and is not - taken into account. The process can be framed by decision schemes (e.g. decisions trees) representing alternative actions. In light of observed conditions, the consequences of the various alternatives can be assessed and even quantified.

PRESENT DROUGHT PREDICTION CAPABILITIES

The current drought prediction capabilities, including efforts being made to monitor drought and to improve our understanding of the physical and dynamical processes responsible for long-term drought. We begin in subsection 3.1 by describing current drought monitoring approaches. From a prediction perspective this can, in part, be considered as "nowcasting", in the sense that the goal is to determine the current state of drought. Next several approaches are described for making hydrologic predictions on seasonal time scales. The last subsection describes several efforts that are relevant to the long-term (multi-year) drought prediction problem, including efforts to assess the impact of climate change.

Drought Monitoring Based on Drought Indices

A simple way to monitor drought operationally is to use drought indices. As already mentioned in the previous section, droughts are typically classified as meteorological (linked to precipitation deficits), agricultural (linked to soil moisture deficits) or hydrological (linked to stream flow or run-off deficits). Therefore, different indices based on precipitation, soil moisture and runoff data are needed to measure different aspects of drought. To construct a reliable drought warning system, the drought classification should not be model dependent. Data sets should be long enough to cover many drought events so the probability distributions are representative. While different indices measure different aspects of droughts, they should be related to one another.

To monitor precipitation deficits, the commonly used index is the Standardized Precipitation Index (SPI). The SPI is based on precipitation alone (Hayes et al. 1999, Mckee et al. 1993, McKee et al. 1995). The SPI was developed to categorize rainfall as a standardized departure with respect to a rainfall probability distribution function; categories range from less than -3 (extremely dry) to more than +3 (extremely wet). It measures the precipitation deficits but it does not take into account the water supply. The advantage is that it covers different time scales and can be applied to station precipitation data. The precipitation data are now readily available in near real time so it is convenient for real time drought monitoring. Sims et al. (2002) suggested that the SPI was more representative of short-term precipitation and a better indicator of soil wetness than the Palmer Drought Severity Index (PDSI).

Another commonly used index is the PDSI, which is based on the water balance between soil moisture supply and demand (Palmer 1965). Categories range from less than -4 (extreme drought) to more than +4 (extreme wet conditions), and have been standardized to facilitate comparisons from region to region. Alley (1984) identified some positive characteristics of the PDSI that contribute to its popularity: (1) it is an internationally recognized index; (2) it provides decision makers with a measurement of the abnormality of recent weather for a region; (3) it provides an opportunity to place current conditions in historical perspective; and (4) it provides spatial and temporal representations of historical droughts. However, there are many shortcomings of the PDSI (Mo and Chelliah 2006; Karl 1983, 1986; Kunkel et al. 2008): (1) it may lag emerging droughts by several months; (2) it is less well suited for mountainous land or areas of frequent climatic extremes; (3) it does not take into account streamflow, lake and reservoir levels, and other long-term hydrologic impacts, such as snowfall and snow cover. Also, many parameters used in the PDSI are sensitive to the calibration procedures.

More recent indices include those based on evapotranspiration (ET) derived from satellite or from a Land Data Assimilation System (LDAS). ET is small during drought, but a small ET does not always imply drought since, for instance, it is typically small during the dry season. Moreover, in humid regions, drought conditions may occur when enhanced ET is taking place. Nevertheless, ET can be used to measure drought development and it is closely related to vegetation cover or type..

In addition to precipitation deficits, the reduction of stream flow, and soil moisture deficits also contribute to economic and agricultural losses associated with drought. As long-term observations of runoff and soil moisture are scarce, the data sets produced by LDAS's become valuable alternatives. The soil moisture and runoff index is based on model-derived soil moisture and runoff as drought indicators; it uses percentiles and the values are normalized from 0 (dry) to 1 (wet conditions). For example, over the United States, Sheffield et al. (2004), Mo (2008), and Andreadis et al. (2005) used products from the North American Land Data Assimilation System (NLDAS) to derive drought indices for drought monitoring. The soil moisture deficit is typically measured by the soil moisture anomaly percentiles. The runoff or streamflow deficit is measured by the percentile of normal of runoff or standardized runoff index or SRI. The SRI is similar to the SPI, but runoff is used instead of precipitation. Most land data assimilation systems have the spatial resolution of 0.5o or higher so the resolution is fine enough for regional applications. Operationally, the meteorological drought based on SPI is defined by SPI values less than -0.8. To classify agricultural drought based on the soil moisture percentiles, the soil moisture percentiles need to be less than 20 per cent (Svoboda et al. 2002, Andreadis et. al. 2005). To classify hydrologic drought based on SRI, indices need to be less than -0.8. Different indices have different time scales. Soil moisture percentiles have the longest time scales and precipitation anomalies have more high frequency components. Roughly, 6-month SPI, and 3-month SRI and soil moisture percentiles have comparable time scales.

The major difficulty in using runoff and soil moisture from the LDAS for drought monitoring and classification is that these indices depend on the model and forcing (precipitation, down-welling short and long wave radiation, surface wind speed, surface air temperature and humidity, and surface pressure). For soil moisture percentiles and SRI, the differences are relatively small among different models driven by the same forcing (Dirmeyer et al 2006a, Robock et al. 2004). However, differences are much larger between models driven by different forcing. For example, the soil moisture percentile differences over the western United States can be as large as 20 per cent: too large for drought classification. Lack of soil moisture observations in most regions of the world prevents the validation of models in this respect. There are also regional uncertainties related to snowfall forcing. For example, Lohmann et al. (2004) found large regional differences (by almost a factor of 4) among the runoff over the United States as estimated from four different NLDAS models.

The major differences in the forcing come from the differences in the precipitation analyses. Over certain regions, the station reports are sparse in real time (e.g. Oki et al. 1999, Koster et al. 2004b, Seneviratne et al. 2010). When precipitation coverage is sparse, the precipitation analyses depend on the method and station reports availability. In real time, the station precipitation reports over the western region of the United States drop to one third in comparison to the historical record, leading to large differences in soil moisture percentiles and runoff.

Hydrologic Drought Prediction at Seasonal Time Scales

Drought indices, such as the SPI, soil moisture percentiles and runoff indices can be derived from the output of global model ensemble forecasts. In general, current global climate models have a horizontal resolution of several hundreds of kilometers (2 degrees latitude/longitude), while the resolution needed by water supply managers and by environmental impact studies is typically 1/8 degree or even higher. Therefore, global model output cannot be used directly for hydrometeorological applications, and some type of downscaling needs to be done. Recently, however, high-resolution global model runs are becoming available, and research is ongoing to assess their skill and whether they can be used for drought monitoring.

While high-resolution global models are a promising approach, most current methods rely on high-resolution hydrology models driven by coarse resolution ensemble global model forecasts. This approach is used because, while the global climate models can provide large-scale circulation anomalies, they are too coarse to provide small-scale hydrologic information like soil moisture and runoff in a given river basin. However, current skill levels for seasonal precipitation and temperature forecasts from global models are still quite modest, so that the contribution of the down-scaled (and error corrected) precipitation and temperature forecasts to the skill of the hydrologic (soil moisture and runoff) forecasts is likewise modest. In this approach to prediction, runoff is routed to calculate streamflow for the major and minor river basins. There are in fact several steps in the process, involving downscaling the global models, running the hydrological models, and post-processing.

Downscaling and Error Correction

There are several commonly used methods for downscaling and error correction. Dynamic downscaling requires a meso-scale regional model. A global model provides the boundary and initial conditions for the regional model. The expectation is that the high-resolution regional model should have better simulations of low level flow anomalies including more realistic moisture transport. It is also expected that the regional model produce better temperature and precipitation on regional grids. Bilinear interpolation includes a correction for any bias in the model climatology. The bias correction and downscaling (BCSD) method is based on probability distribution functions (Wood et al. 2002, 2005; Wood and Schaake 2008). If a variable is normally distributed, this method corrects both the mean and standard deviation. The BCSD method does not takes into account the hindcast skill of the global model forecasts.

Linear regression, in addition to correcting the mean and standard deviation, weights the downscaled and error corrected field by the correlation between the hindcasts and observations (Wood and Schaake 2008, Schaake et al. 2007). Bayesian merging and error correction is based on a Bayesian framework (Luo et al. 2007, Luo and Wood 2007,2008). The climatological distribution of a variable serves as the prior. Then, a linear regression between the global climate model hindcasts with all members and the corresponding observations is used to derive the likelihood function. The posterior distribution is the merged forecast. This method corrects the mean, standard deviation and takes into account the post hindcast skill and the spread of the forecasts. The Constructed analogues method assumes that if one could find an exact analogue of today's field in the historical record, then that field in the future should replicate the evolution in time of that exact analogue. In practice, the analogues are constructed by linear regressions (Maurer and Hidalgo 2008, Hidalgo et al. 2008).

The Hydrologic Forecasts

There are currently several approaches to hydrologic forecasting. These are outlined below. Note that only the last approach takes advantage of downscaled global model forecasts of precipitation and temperature.

- Traditional Ensemble Streamflow Prediction (ESP): The hydrologic forecast requires a hydrology model. The most common used models are the Variable Infiltration Capacity Model (VIC) model and Sacramento SAC model. We note that most of the NWS Regional Forecast Centers use the SAC model while NCEP used the Noah LSM. The model needs to spin up to the date that the forecast will be made. A year and month are randomly selected. The daily time series of precipitation and temperature from that month and year are used to drive the hydrology model to produce soil moisture and runoff forecasts. The forecast skill comes mainly from the initial conditions.
- Conditional forecast for ENSO years: This is similar to the above traditional forecast, but ENSO years are selected rather than random years. The selection can be keyed to different climate indices in addition to ENSO.
- Forecasts based on bias corrected and downscales global forecasts. The hydrology model is spun up to the initial condition date. The daily precipitation and temperature are derived from the monthly ensemble mean of bias corrected and downscaled global forecasts. They are scaled so that the monthly total precipitation and temperature are equal to the bias corrected precipitation and temperature from the climate forecast. These forecast variables are used to drive the hydrology model to produce soil moisture and runoff forecasts. The forecast skill comes potentially from both the initial conditions and from the global forecasts.

Post-Processing and Skill Assessment

There are many hydrologic runs that can be made by using different initial conditions or different forcings. Post processing is needed to produce ensemble mean anomaly streamflow and soil moisture forecasts. The soil moisture percentiles and streamflow percentiles can be derived from the ensemble members. Overall, the hydrologic forecasts are skillful for lead times of 1-2 months. In snow dominated hydrologic regimes (e.g. western North America), knowing the snow pack provides considerable skill through the melt season but the lead skillful times fall off after the melt season. The skill comes from initial conditions and global forecasts. During ENSO years, the global forecasts are overall more skillful. In that case, skill for the hydrologic forecasts is also higher. If Bayesian error correction or linear regression methods are used to downscale and correct errors, then these methods take into consideration the hindcast skill. If the precipitation and temperature forecasts from the global model are not skillful, then these methods will use climatological forecasts for precipitation and temperature, and then the skill is close to that of the ESP forecasts.

Wood and Lettenmaier (2009) provide an overall assessment of current hydrolic seasonal forecast skill in the extratropics stating that "... in our experience, most of the actual land hydrologic (and hence water resources) seasonal prediction skill comes not from an ability to forecast climate, but rather from hydrologic initial conditions. These include knowledge of soil moisture and snow water storage. Furthermore, this skill is often at the shorter end of the seasonal time scale (e.g., weeks)".

Efforts to improve the utilization of current and future (e.g., satellite-based) observations of soil moisture and snow for stream flow forecasts are on-going.

MULTI-YEAR DROUGHT PREDICTIONS AND PROJECTIONS

The prospects for predicting drought on multi-year to decadal time scales are as yet unclear. To a large extent these hinge on improvements in our coupled models and on addressing basic predictability issues concerning the relevant SST anomalies and links to regional drought. While some efforts are underway to carry out experimental decadal predictions, much of the current work is focused on improving our understanding of the basic physical and dynamical mechanisms and phenomena that operate on multi-year time scales and that could potentially provide some skill in predicting regional drought on these time scales. Efforts to assess possible changes in drought characteristics under global warming scenarios are further limited by uncertainties in how the various components of the atmospheric hydrological cycle (e.g., clouds, precipitation, evaporation) respond to a warmer climate, the role of aerosols, vegetation changes, as well as a basic lack of understanding of how the relevant low frequency coupled (atmosphere/ocean) variability might be altered in a warmer climate.

The U. S. CLIVAR Drought Attribution Experiments

A set of GCM experiments were designed by a U. S. CLIVAR drought working group to study the relationship between oceanic SST forcing and land surface feedback on drought. This was a multi –institution coordinated effort involving five different AGCMs and one coupled model. The models were forced by prescribed SST anomalies consisting of the leading patterns of annual mean SST variability. The project and some of the initial findings are summarized in Schubert et al. (2009).

One of the key findings of the USCIVAR working group study was that all the AGCMs produce broadly similar precipitation responses to the leading Pacific SST forcing pattern, with a cold Pacific leading to reduced precipitation and a warm Pacific leading to enhanced precipitation over most of the United States. While the response to the Atlantic pattern is less robust, there was general agreement among the models that the largest precipitation response over the U.S. tends to occur when the two oceans have anomalies of opposite sign. The response to the positive SST trend forcing pattern is an overall surface warming over the world's land areas with substantial regional variations that were in part reproduced in runs forced with a globally uniform SST trend forcing. The precipitation response to the trend forcing, on the other hand, is weak in all the models.

The major drawback of these experiments is the implicit assumption that the SST modes in different oceans are independent and do not interact, even though it is known for example that SSTAs in the north tropical Atlantic can be influenced by ENSO. Nevertheless, these results clearly demonstrate that skillful predictions of long-term drought in many regions throughout the world depend critically on being able to predict the leading low-frequency patterns of Pacific and Atlantic SST variability. The results also highlighted the importance of, and uncertainties (model differences) in, the land/atmosphere feedbacks.

DRICOMP and other Relevant IPCC Studies

A number of studies have provided comprehensive analyses of agricultural and hydrological droughts in CMIP3 simulations. Wang 2005) examined the impact of greenhouse gas warming on soil moisture based on predictions of 15 global climate models. They found considerable inter-model differences in the sensitivity of the soil moisture response and suggested that a major part of the differences are attributable to differences in the land surface parameterizations.

Nohara et al. (2006) investigated the projections of river discharge from 24 rivers during the twenty-first century as simulated by 19 AOGCM simulations based on the SRES A1B scenario. They found that although individual models have difficulty reproducing present river discharges, a weighted ensemble mean of the models produces more accurate discharges for most rivers, except those affected by irrigation and dams and evaporation from the river surface. Sheffield and Wood [2008] analyzed projected changes in drought occurrence under future global warming in CMIP3 simulations and found large increases in long-term drought frequencies in the

Mediterranean, west African, central Asian and central American regions. To estimate regional impacts on hydrology, several studies have used downscaled precipitation and temperature from IPCC-type experiments to drive a hydrology model such as VIC as described above. The changes of runoff and ground water are estimated from the output of the hydrology model. It was found, for example, that the impact on hydrology is not limited to the annual mean precipitation or runoff. Studies of the Colorado River (e.g., Christensen et al. 2004, Payne et al, 2004) indicate that the annual mean river flow does not change much, but seasonal flows shift toward larger winter and spring flow and smaller summer and autumn flow. This implies that there will be less water available in summer especially in low flow years.

The IPCC (2007) report also documents that the subtropical dry areas of the word will become drier in association to a weakening and poleward explansion of the Hadley circulation. According to Lu et al. (2007), the expansion of the Hadley cell is caused by an increase in the subtropical static stability, which pushes poleward the baroclinic instability zone and hence the outer boundary of the Hadley cell. Thus, the climate models need to correctly simulate the dynamics associated with the Hadley circulation.

ENSO also plays a key role in drought in a number of regions throughout the world, yet the simulation of ENSO in IPCC-class coupled models remains a challenge.

The U. S. CLIVAR drought in coupled models project (DRICOMP) specifically focused on the use of coupled model experiments to assess whether coupled models are able to simulate drought and to quantify the roles of the oceans, the seasonal cycle and land surface feedback in drought. The impacts of drought on water availability and the distinctions between drought and drying were also examined. The overall objective was to encourage diagnostic research into the physical mechanisms of drought and evaluate whether the current models are able to simulate those mechanisms. The project encouraged the analysis of the results of the WCRP CMIP3 simulations.

For example, Caio and Goddard (2009) found that the subset of CMIP3 models with realistic ENSO variability do appear to produce realistic 20th century drought variability in the tropics and do not predict robust changes in the strength or frequency of El Nino events or in the drought related teleconnections in the twenty-first century. Capotondi and Alexander (2009) examined six IPCC-AR4 models and found statistically significant correlations between decadal precipitation anomalies in the Great Plains region and tropical Pacific SSTs in all the models. However, the influence of other oceans (Indian, tropical and North Atlantic), which previous studies have identified as potentially important, appear to be model dependent.

In contrast to projections for many tropical regions, there is a high degree of consistency among models regarding projected sub-tropical precipitation changes. Mechanisms include a general increase in moisture divergence across the sub-tropical dry zones, a result of mean climatological moisture divergence and rising humidity (Held and Soden, 2006; Lu et al., 2007; Seager et al., 2007); the northward expansion of the Atlantic Hadley Cell and to a large extent a positive trend of the Northern Hemisphere Annular Mode (Previdi and Liepert 2007). As part of the sub-tropical drying, Giorgi 2006 has identified the Mediterranean region as a "Hot Spot" in hydrological change, with a decrease in mean precipitation and increase in precipitation variability. Mariotti et al., 2008 found that in CMIP3 regional drying tendencies are exacerbated by warming enhanced evaporation.

McCrary and Randall (2009) found that long-term droughts in the US Great Plains in three AR4 model simulations of the 20th century are comparable in duration, severity and spatial extent to the observed ones. However the processes found to be associated with simulated long-term droughts vary among the models. In both CM 2.0 and HadCM3, low-frequency variations in Great Plains precipitation are found to be associated with low-frequency variations in tropical Pacific SSTs. In CCSM3, on the other hand, there appears to be no significant correlation between tropical Pacific SST variability and Great Plains precipitation. Instead, strong land-atmosphere coupling may explain the persistence of long-term droughts in this particular model. Biasutti et al. (2009) found that projections of Sahel rainfall in the 21st century differ greatly among CMIP3 models. The discrepancies appear to be linked to changes in the Saharan low which suggests the need to better understand the mechanisms driving the variability of the Saharan low. Lyon (2009) found that CMIP3 models exhibited less ability in reproducing the observed conditional probability of a heat wave in Southern Africa under El Niño conditions. This effect appears to be related to a comparatively weak seasonal precipitation teleconnection pattern into southern Africa in the models during El Nino when drought conditions often develop. The heat wave–drought relationship did not substantially change in climate projections when computing anomalies from future climate means.

GLACE 2 Experiments

The GLACE2 international effort to understand the role of land initialization on seasonal predictability/prediction should also be mentioned.

RESEARCH AND DEVELOPMENT NEEDS

As already emphasized in the previous sections, drought is typically defined in terms of its links to meteorology, agriculture, hydrology and society. This acknowledges the broad array of physical processes, time scales, and impacts on human activities that together determine the nature and severity of droughts. As such, advancing our ability to predict drought requires, at least in principle, improvements in modeling and observations encompassing all components of the physical climate system (atmosphere, hydrosphere, cryosphere, land surface, and biosphere) as well as improved knowledge of the impacts and interactions with human activities. While progress is being made in all these areas, major advances in the drought prediction problem will almost certainly require a more focused but still interdisciplinary program that builds on our current understanding of the nature of drought and encourages work on those aspects and variables that together link the evolution of drought from its inception (the drivers, if they exist) to the eventual impact on quantities that most directly affect human activities. In the following we outline the research and development needs for a program specifically geared to improving our ability to predict drought with a focus on the three key variables of precipitation (most relevant to meteorological drought), soil moisture (most relevant to agricultural drought) and stream flow (most relevant to hydrological drought).

We may consider precipitation as the primary variable in terms of addressing the underlying causes and predictability of drought (with soil moisture and stream flow acting as secondary variables that can, through initialization and feed backs, act to enhance drought prediction skill). On the other hand, in terms of impacts on human activities, it is likely that soil moisture and stream flow are of equal if not greater importance. In that way, maximizing the relevance to society requires that precipitation not be studied in isolation, but be considered in the context of the "pathway of information" that leads to predictability in soil moisture and stream flow (Fig. 3).

Precipitation

Precipitation is a key variable in the drought problem, and it is of course a key part of the global hydrological cycle. It has proven useful in characterizing drought. In fact experts participating in a recent workshop on Indices and Early Warning Systems for Drought (WMO 2009) agreed that all National Meteorological and Hydrological Services around the world should use the Standardized Precipitation Index or SPI to characterize meteorological droughts.

Precipitation can be viewed as the active ingredient that links remote SST anomalies to regional changes in hydrology. Numerous studies have now shown that SST anomalies (especially tropical anomalies) force convective (and therefore heating) anomalies that generate global-scale circulation anomalies that in turn impact regional climates throughout the world. The impact on regional climates is fundamentally a weather/climate problem and requires models with high spatial resolution to accurately simulate weather-scale phenomena and ultimately resolve terrain features on catchment scales relevant to simulating the impacts and feedbacks associated with snow, soil moisture and stream flow.

In light of the above discussion it is clear that any major advances in the prediction of precipitation anomalies on seasonal and longer time scales is largely dependent on improvements in our coupled atmosphere/ocean models and their ability to simulate the key SST signals that drive changes in atmospheric circulations that lead to regional drought. This includes the simulation of the SST anomalies associated with ENSO, the PDO, the AMO, and trends such as those associated with global warming. Improved models would allow estimating basic predictability, and when combined with improved ocean assimilation and reanalyses, could lead to skillful predictions of the leading drought-related SST anomaly patterns. The links between low frequency variability and climate change are under active investigation. In fact, the extent to which global warming will impact drought directly (e.g., by altering the meridional extent of the Hadley circulation, Lu et al. 2007) or by altering the characteristics of low frequency variability (ENSO, PDO, AMO, NAO or AO, etc., which in turn impacts drought) is still poorly understood, and requires more basic research. The current efforts at doing decadal predictions in support of the next IPCC report should provide some basic insights into predictability at these long time scales. The land also plays a potentially key role both in terms of the feedbacks between soil moisture and the atmosphere, the impact of aerosols, and changes in vegetation. More work is needed to better quantify the basic land/atmosphere coupling strength, and to include more accurate treatments of aerosols and vegetation.

On shorter (subseasonal) time scales, one perhaps needs to distinguish between droughts and "dry spells" with the latter characterized by relatively short (weeks to a month or two) periods of reduced precipitation and high heat but with less impact on longer term agricultural and hydrological drought conditions. Nevertheless, instances of reduced precipitation (and excessive heat) lasting one or two months can have devastating societal and ecosystem impacts as evidenced by the 2003 European heat wave, and the well-known occurrence of Indian monsoon breaks that can be a determining factor as to whether a particular monsoon season will be good or bad. On these time scales, the ability of the atmospheric general circulation models to simulate and predict basic subseasonal and weather variability plays an important role. This includes phenomena such as the Madden Julian Oscillation (MJO), tropical storms, middle latitude weather systems, and other low frequency subseasonal variability such as the Pacific North American Pattern (PNA), North Atlantic Oscillation (NAO – though we note that the NAO has sub-seasonal to decadal variability characteristics), Southern Annular Mode (SAM), and the Northern Annular Mode (NAM). At these time scales, coupling with the land also plays a potentially important predictive role.

Specific model needs include:

- Improved simulations of SST climate, seasonal cycle, and variability on seasonal to decadal and longer time scales in coupled atmosphere/ocean models (including any regional variations in SST trends associated with global warming)

- Improved global and regional interactions of the atmosphere with SST including regions and time scales at which the atmosphere is an important driver of ocean variability.
- Improved response of the atmosphere to SST forcing including the seasonal dependence of the responses in planetary waves, the diabatic heat sources and sinks at middle-upper troposphere and the stratosphere, storm tracks, Hadley and Walker circulations, and regional moisture fluxes that contribute to changes in precipitation.
- Increased horizontal and vertical model resolution to provide more realistic weather/climate linkages (e.g., MJO, tropical storms) and other impacts on regional-scale climate phenomena (e.g., LLJs)
- Improvements in land-atmosphere coupling
- Improved treatment of aerosols
- Improved treatment of vegetation
- High-resolution global model forecasts have already been used successfully for hurricane forecasts in the Atlantic and eastern Pacific. Diagnostic studies and research are needed to assess their skill in hydrologic forecasts and usefulness in drought monitoring.
- Research on integration of satellite-based precipitation estimates in coupled land-atmosphere models.

In addition to model improvements there is a need for further assessments of precipitation variability and the characterization of meteorological drought from both observations and model simulations by taking advantage of existing drought indices (e.g., SPI) and encompassing diagnostic and prediction tools derived from both tropospheric and stratospheric parameters.

Traditionally, the SPI is derived using long-term rain-gauge measurements. However, rain gauges are sparsely distributed in space, and thus do not necessarily provide a reliable areal representation of precipitation. This can be a major limitation for various drought-related analyses, including estimation of the spatial extent of droughts, deriving probability of occurrence of extreme droughts and their spatial dependencies. In addition to rain-gauges, weather radar systems provide high-resolution precipitation data. However, the density and coverage of current weather radar and gauge networks are inadequate for monitoring global precipitation.

The limitations of weather radar systems and rain gauge stations highlight the significance of satellite-retrieved precipitation data for large-scale drought analysis. In the past two decades, various satellite-derived precipitation algorithms have been developed to retrieve estimates of precipitation in global scale. Currently, satellite-based precipitation data are actively being used for drought monitoring and analysis. For example, Using TRMM and PERSIANN data sets, the experimental African Drought Monitor provides near real-time monitoring of drought conditions. Due to the potential significance of satellite-based precipitation data to current and future drought research, there exists a critical need to improve the available global precipitation retrieval algorithms. In addition to the satellite estimates, the new generation of reanalyses are beginning to provide useful and dynamically consistent precipitation estimates that allow diagnostic studies of the complete water and energy budgets

Relevant research areas include: interannual to decadal and longer changes in the water and energy cycles; decadal ocean variability in the context of regional drought; connecting ENSO and other shorter-term SST variability to the initiation and demise of long-term drought.

Specific diagnostic/analysis and observational needs include:

- Improved long-term (multi-decadal) atmosphere, ocean and land reanalyses for initialization, calibration and validation of forecasts. The reanalyses must be able to take advantage of a host of existing and new measurements describing the surface and subsurface oceans including ocean biology, atmospheric chemistry and aerosols, and variations of the subsurface land water storage (e.g., NASA's GRACE and possible higher resolution measurements on future missions). New data assimilation methodologies need to be developed that take into account the inhomogeneities (both temporal and spatial) in the global observing system.
- Improved satellite-based global precipitation retrieval algorithms.
- Improved high resolution multi-sensor merged precipitation products (e.g., multi-satellite-derived gauge-adjusted)
- Uncertainty estimates of global precipitation data as it plays a significant role in uncertainty of drought analysis.
- Improved adaptive data assimilation techniques in order to derive the best estimates of area average precipitation at any given location using all available data sources (gauges, radars, satellites).
- Improved global assessments of the nature (mechanisms and types) of precipitation variability and extremes from station observations, reanalyses, and paleo data. This includes efforts to characterize the spatial and temporal variability (including drought migration), the seasonality, and the signal to noise ratios associated with global scale SST forcing (e.g., ENSO, PDO etc). It also includes documenting historical droughts, assessments of basic regional climatology including precipitation type (frontal, convective, monsoonal), and the use of various tropospheric and stratospheric indices.
- In addition to improving the skill levels, obtaining realistic measures of basic predictability would help to quantify the gains possible gains in forecast skill with improved models and initial conditions
- Research on spatial dependencies of extreme droughts (multivariate analysis drought severity, duration and spatial extent).

In addition, drought prediction would benefit come from improvements of prediction methodologies such as improved initialization techniques, ensemble generation and multi-model approaches.

Soil Moisture

Soil moisture potentially provides a predictive signal through memory of the initial conditions and through feedbacks that can act to enhance initial precipitation anomalies. It is also the quantity of key importance to the agricultural community, impacting the growth of crops and other vegetation. Recent research shows that there are distinct regions where land-atmosphere interactions seem to be especially important. Land impacts and associated potential predictability can be considered to have two components. These consist of 1) the memory inherent in the reservoir holding capacity of water in the soil that interacts with the atmosphere, and 2) the land/atmosphere coupling associated with the feedback between the soil moisture and precipitation.

There are few soil moisture measurements so that most soil moisture information currently used for the initialization of climate models is based on the output of land models forced with observed precipitation and radiation fields.). Recently a global soil moisture network has been established which collects the in situ soil moisture measurements of existing networks and includes the Global Soil Moisture Databank data.

Validation datasets remain problematic for soil moisture. There are few soil moisture measurements so that most soil moisture information currently used for the initialization of climate models is based on the output of land models forced with observed precipitation and radiation fields.

Soil moisture estimates are also available from remote sensing retrievals, but these have several limitations: Microwave remote sensing estimates only capture soil moisture anomalies in the top few centimeters of the soil, and cannot provide measurements in regions with dense vegetation cover, and measurements from the Gravity Recovery and Climate Experiment (GRACE) do not distinguish between soil moisture, ground water, snow cover and surface water, and have a very coarse horizontal resolution (several 100 km; e.g. Tapley et al. 2004, Ramilien et al. 2008). In the future, measurements provided by ESA's SMOS satellite (launched in 2009) and NASA's SMAP satellite will provide improved microwave remote sensing measurements of surface soil moisture, but still be of limited use for estimating root zone soil moisture.

While soil moisture is the key variable of interest to the agricultural community, it cannot be considered in isolation when addressing the predictability of growing season soil moisture. In particular, snow pack and temperature play key roles in determining both the magnitude and timing of the spring melt - a key source of soil moisture in certain climates.

Specific issues that need to be addressed include:

- Uncertainties in land-atmosphere coupling (Koster et al 2004, Dirmeyer et al. 2006b, Orlowsky and Seneviratne 2010)Land Impacts has two components (need to reduce uncertainties in both):
  - Memory issues: what is the reservoir holding capacity of water in the soil that interacts with the atmosphere, deep soil memory.
  - Feedback issues: model uncertainty about strength of feedbacks.
- Must consider ties to snow and temperature (for instance, timing of spring melt or total snow amount; e.g. Hall et al. 2008). Improved soil moisture forecasts will require more skillful P and T forecasts produced by global models
- What is the role of vegetation on longer (multi-year) time scales? Models are still crude.
- Role of aerosols, dust? -improved modeling of aerosols and impacts in global models (resolution issues, link to clouds), e.g., Zeng (2003)
- Connect to applications models. Assess needs in terms of time scales, resolution, quantities (e.g., run-off, evaporation)
- Observations: global soil moisture estimates (e.g., from the current ESA SMOS and planned NASA SMAP missions), soil moisture profiles (for monitoring "sensitive" regions, such as the Great Plains), snowpack, land model forcing data (precipitation, radiation, etc.) for land data assimilation, vegetation properties (e.g., from NDVI, EDVI MODIS data), surface temperature, and streamflow. These observations are important for both prediction initialization and verification.

Stream flow

Improved forecasts of stream flow will require the following:

- improved initial conditions of soil moisture (including under the snow) and snow pack; also need to include the assimilation of observed river discharges (only possible if SWOT mission moves forward)
- increase the number of surface observations (monitoring)
- improved high resolution and/or downscaled forecasts of precipitation and temperature to drive stream flow models (down-scaling temperature/precipitation linkages/snow melt). In addition to improving individual models, this should include multi method and multi model ensemble forecasts,
- reduced uncertainties in surface/subsurface water reservoirs (including estimates of recovery time).
- characterization of the links to drought regions and issues ("upstream" river basins, seasonality, snow, glacial melt, soil moisture, run-off, spring melt, frost, time scales).
- addressing issues with observations (need to remove signature of

human controls). Incorporating multiple land surface models as well as water reservoir storage in the river routing scheme;

- a post-processor that removes hydrologic model bias in the hydrologic forecast products;
- improved extreme event prediction, including characterizing extreme events.
- Improved verification of streamflow forecasts. For the United States they can be verified against the streamflow data from the USGS. For soil moisture, the difficulty is that there are not many measurements available. One could evaluate soil moisture forecasts against soil moisture simulations from the NLDAS, even though one recognizes the uncertainties in the NLDAS.

Findings and Recommendations

In summarizing the report and organizing our recommendations we divide this section into four elements, addressing major research gaps, priorities, coordination, and how to best leveraging WCRP activities. We begin with research gaps. Major research gaps and the need for an integrated approach to drought prediction. Here we consider those research gaps that, if filled, would advance the understanding of the drought processes and drought prediction:

When looking at the phenomena of drought it appears customary to look at the various processes and variables involved first while if one looks at the prediction of drought a specific feature becomes more important which is the timescale. Hence, this dealt with specifically.

Features of Drought

Drought is a persistent lack of precipitation although this does not have to be in the same area where the drought is occurring. Hence, it is critical to focus considerable attention on the 'precipitation' part of drought. This has in general been a major research 'gap' with most of the attention being paid to large scales and to surface issues. These are obviously crucial but so are precipitation issues.

What mechanisms are operating solely or in combination to sustain the lack of precipitation? There are many possibilities from large scale conditions down to microphysical factors. It only takes one mechanism at a time to ensure that no precipitation reaches the surface. (Droughts in Canada for example are linked with both cold as well as warm temperature anomalies and these in turn can be linked with different means of shutting off precipitation production. Droughts can be nearly cloud-free or they have many clouds, just not capable of producing large amounts of precipitation. As well, droughts can be windy to generate dust and aerosols which may alter precipitation production or they can be calm with no dust.) What are the chains of events associated with droughts? The detailed nature of drought is, to some degree, an integration of events that happened previously. To what extent do drought areas interact with nearby wet areas to enhance and/or reduce gradients between them? Under what conditions does the dry air over a drought region lead to enhanced precipitation through alteration of the character of precipitation systems? How do droughts end? To what extent are there common and/or unique chains-of-events?

What critical thresholds exist for precipitation aloft to not reach the surface? Virga is common in at least some drought situations but what are the precise conditions for this as opposed to precipitation on the ground?

Specific needs to improve drought monitoring include:

- Improved precipitation analyses and data coverage.
- Establishment of real time soil moisture and streamflow networks: these data would be valuable for model evaluation as well as drought monitoring.
- Better flux measurements. For example: flux measurements from the Ameriflux network can be used to validate the NLDAS products and improve model parameterization.

Time Scales of Drought

The research needs for better anticipating droughts concerns basically 3 time scales:

1. *Subseasonal to Seasonal time scale:* Are concerned summertime droughts which could be better anticipated by early warning systems. The main predictability potential lies, in the initial condition of a forecast, in the SST anomalies, soil moisture anomalies and snow cover and depth anomalies (for spring season). In order to be able to use this potential, accurate global-coverage observations of these quantities must be available. SST anomalies are usually correctly described by observing systems, but such is not the case of soil moisture and snow depth.
   - From the observational standpoint, the improvement of seasonal forecasts (either dynamical or statistical) requires a better 3-dimensional description of soil moisture and its spatial variability. It also requires an accurate description of the vegetation state and the land cover. Priorities could be to reinforce/ develop multi-scale networks of SM observations (nationally) and networks of networks, and to build soil moisture data assimilation systems using both satellite and in-situ observations.
   - Predictability also relies on the ability of models to faithfully reproduce the key processes and feedbacks leading to and maintaining droughts.
     - The large-scale dynamical response to SST anomalies is a fairly well documented domain, but we need to better

understand the divergence among the models responses. More model intercomparisons are needed to quantify these responses.

- The soil moisture/ energy budget/ cloud/ radiation interactions and feedbacks are a weak point in models. Here also more model intercomparisons (surface energy budget, clouds, feedbacks) may be beneficial.
- It is also important to develop statistical methods to quantify relations between soil moisture, precipitation, clouds,radiation, weather patterns, ...: that is to say, time series analyses of correlations between these variables, composite analysis, or more advanced methods to identify spatial patterns of correlations. Observational statistics should be compared also with equivalent model statistics to pinpoint model deficiencies.

• Successful drought seasonal prediction also relies on our ability to evaluate uncertainty and to use the huge potential offered by the international ensembles of models. Advanced statistical methods to provide probabilistic forecasts and their evaluation against observations or analyses should be developed and further refined.

At this seasonal scale coordinated efforts on hindcasts of a few extreme cases, with extended diagnostics should be recommended.

2. *Decadal time scale:* Decadal prediction of droughts relies basically on
   - The ability of models and their ability to represent the correct SST-dynamics relations and the soil moisture/ cloud/radiation interaction processes.
   - The ability of coupled models to reproduce interdecadal variability
   - Our ability to describe ocean initial conditions.

The corresponding research needs are the same as the decadal projection needs. Efforts on downscaling techniques (statistical/dynamical) could further improve regional focus on hydrology (at catchment scale).

It is to be emphasized that essentially at this time scale hindcasts will not be of much help. They should however help to identify climate model drifts.

3. *Climatic (multi-decadal time scale):* At the time scale of 50 years or more, all the above mentioned questions (processes, ocean and land surface coupling, interdecadal variability) are all essential, and previous remarks apply. A new key question is whether projected extreme events have well predicted statistics by models, as concerns their amplitude, frequency, impacts, ... This can only be tackled on a statistical basis. The climate prospective therefore calls for an increase in diagnostic studies. It is essential to develop

- A better knowledge of drought and heat waves from past data, including paleoclimatic data from direct and indirect observations, in order to better document the statistics and characteristics of droughts in natural climate variability;
- An interdisciplinary knowledge from recent (and current) regional case studies, including all the socio-economic and ecosystemic impacts, as well as interactions with carbon cycle;
- Intercomparisons on model statistics of droughts and drought-related processes, including the ability of models to simulated relevant dynamical weather regimes, precipitation frequency and intensity, cloud, radiation etc...

Program priorities

Making the effort to fully document drought with all its subtlety (data integration but with a specific purpose)

Focusing more on the structure of drought. For many countries without huge resources, it is a way for them to contribute. This in general would highlight 'regional and local' issues. The national scale is important for adaptation policy but generally droughts have regional characteristics and impacts (a scale that is usually larger than a single nation). It could be interesting

- To identify a few main regions of the world (link with CORDEX strategy?) with interest to neighboring countries for impacts and to international research for earth-system functioning and modeling,
- To develop an international case-study approach, which would include
  - developing shared interdisciplinary databases on each case (physics, biogeochemistry, socio-economical impacts);
  - performing regional climate model inter-comparisons;
  - performing seasonal hindcast intercomparisons;
  - projecting expected occurrence of the event type in the future climate;
  - Diagnosing past analogues from reanalyses and paleoclimate data;
- To develop a "research-oriented watch" on major drought cases as they appear, by
  - coordinating and follow research by specific workshops on regional cases as they appear, following the event
  - developing an interdisciplinary international approach on these cases by workshops on impacts
  - develop a portal gathering all seasonal drought predictions in each regions
- To develop a kind of "DMIP" approach (Drought Model Inter-comparison Program) for inter-comparisons. That would include:

- Seasonal hindcasts
- Seasonal forecasts
- Processes diagnostics
- socio-economical and ecosystemic impacts

National and International Coordination

The World Climate Research Programme is in a unique position to stimulate and facilitate national and international research collaboration in the area of drought. A suite of activities carried forward by specific champions can best foster the coordination of these activities. This will ensure a bottom-up approach that is carried and intellectually supported by the community.

DROUGHT ENVIRONMENTS

Hydroponics

Hydroponic culture allows plant cultivation under defined and reproducible osmotic stress conditions by adding an osmoticum such as polyethylene glycol or sorbitol to the culture medium. This system supports testing of a large number of plants for adaptation and/or acclimation of vegetative parts to osmotic stress conditions. However, the artificial experimental setting imposes different stress conditions to the plant than drought in the field, consequently the transferability of findings to natural environments is limited. Further, hydroponics does not allow yield assessment, which is a crucial component in drought tolerance testing of potato. Therefore, hydroponics might provide a suitable testing system for specific adaptation or acclimation reactions of plants to water stress, but cannot replace phenotyping in more natural settings.

Growth Chamber

Growth chambers equipped with temperature and air humidity control are ideal environments for drought tolerance studies providing reproducible drought conditions. But due to space limitation this test system is only practical for a small number of genotypes. Yield assessment is possible in growth chambers, when potato plants are grown in large containers instead of in pots. However, several physiological parameters can be only performed on pot plants, such as transpiration efficiency studies.

Greenhouse

Greenhouse-grown plants provide a suitable experimental setting for many drought tolerance-related studies. As already stated above, for yield assessment plants have to be grown in large containers instead of in pots. For pot-grown plants it is important to avoid position effects by randomly moving the pots. A further concern for greenhouse trials is the amount of light available for the plants, as glass roofs might absorb a significant portion of sunlight and thus cause a strong decrease of irradiation, which might influence drought responses.

Rain-Out Zone

Rain-our zones provide an ideal environment for drought screenings. They consist of a roof protecting the plants from rainwater and are equipped with plastic barriers at the borders and below the test plot to avoid water inflow from the sides or from below ground. Moderate construction costs allow large dimensioned rain-out zones harboring thousands of plants. Increased heat or air humidity due to roofing might influence the drought environment. So either mobile roofs that cover the plants only during a precipitation event, or good aeration of the rain out zone is required.

The surface of rain-out zones should be completely even to avoid water accumulation in pits. The soil depth above the plastic barrier should be at least 50 cm to allow deep rooting. Water should be applied by drip irrigation.

Field

Field testing is the cheapest and the most natural way for drought phenotyping, but also the most affected by environmental impacts. Field testing is relatively straight forward in zones with negligible precipitation, like in dry seasons or desert zones, as long as excessive heat does not interfere with the drought trial. Field plots should have homogenous soil type, should be flat to avoid water accumulation in pits and should be equipped with an appropriate irrigation system, ideally with drip irrigation.

EXPERIMENTAL DESIGN

Generally drought trials are performed using two regimes: a well watered control, which is irrigated to field capacity, and a drought treatment, where plants are exposed to water stress. An appropriate number of repetitions, each with a suitable number of plants, has to be planted for each treatment. In field- and rain-out zone trials, a minimum of 3 blocs of 5 plants per treatment i.e. 3 biological repetitions should be used. When destructive sampling is required, a higher number of plants per bloc is necessary. When space and seed is not limiting, at least 4 repeated blocs of 10 plants should be used for each treatment. The plants at the border of the blocs are not used for evaluations because of influences by neighboring genotypes or field borders. Blocs should be arranged in a random complete bloc design.

In trials with pot plants, a minimum of 5 individual plants per genotype and treatment should be used. Each individual plant represents a biological repetition. The position of the pots has to be completely randomized and border effects should be excluded by randomly moving the plants every few days.

In all environments listened above, except hydroponics, drought might be applied by stopping irrigation at an appropriate time point. Generally drought treatments are applied during a defined developmental stage of the plants, such as tuberization. For rain-out zones or field settings it is recommended to choose clones with similar phenology in order to expose the plants to drought at identical developmental stages. Drought trials using pot-grown plants allow more flexibility, as irrigation can be stopped individually for each pot. Duration of the drought treatment depends of the aim of the study and might be chosen either for a fixed period, dependent of the soil type and irradiation, lasting for 30 to 60 days, or drought might be applied until a certain plant- or soil parameter reaches a defined threshold. The threshold values given in table were applied in potato drought tolerance trials at CIP or by collaborators and might vary with experimental setting, soil type, environment and genotypes.

Parameter	Threshold Value	Evaluation Method	Required Measurements
Soil water content	20%*	Gravimetrical soil water determination	Soil fresh weight Soil dry weight
Soil water potential	-0.6 to -1 MPa	Sensor	Reading
Leaf water content	65-70%	Tourneux et al. (2003)	Weight of fully hydrated leaf Leaf fresh weigh Leaf dry weight
Leaf water potential	–1 to –2.5 MPa	Tourneux et al. (2003),	Callster et al. (2006) Schoellander bomb, osmometry
Photosynthesis efficiency	1 $\mu mCO_2/m^2/sec$	Photosynthesis analyzer	Reading

Tab. Useful parameters to determine duration of drought treatments in phenotyping studies. *) value might strongly vary between different soil types.

An alternative drought treatment involves irrigation with a limited amount of water during the whole cropping season, instead of stopping irrigation completely. The amount of water required to reach stress conditions has to be determined by quantification of evapotranspiration in the given environment. For stress treatments irrigation should be lowered to 30 - 50 per cent of the water evapotranspiration of well watered plants. Under low irradiation condition, providing 200 to 300 mm of water during the whole growing season distributed over weekly irrigation events is a suitable drought treatment for potato.

DROUGHT PHENOTYPING PARAMETERS

Phenotyping has to capture three aspects: the strength of drought, the effect of drought on the plant performance and the plant response. Evaluation parameters for drought trials have to be chosen appropriately to quantify the impact of a defined drought event on plant performance and to distinguish between responses leading to drought tolerance and other stress responses pinpointing susceptibility of the plant to water stress. Environmental data are recorded over the whole duration of the phenotyping experiment and comprise climate data (temperature, irradiation, for field trials also precipitation) and soil water content or potential. Monitoring plant performance involves visual scoring as well as quantitative measurements. Quantitative measurements are taken from at least 3 biological repetitions. Additionally technical repeats of measurements have to be taken to allow determination of means and standard deviations and to assess significance of differences between clones and treatments by T-test. Analysis of genomics data, particularly microarray data, need specific significance analysis beyond T-tests.

MEASURING AND SAMPLING

An experimental bloc consists of at least 5 plants of a clone. For measuring and sampling, the 3 central plants of a bloc are used to avoid neighbor effects. Porometry, spectroradiometry, SPAD and photosynthesis efficiency measurements are performed on the apical leaflet of the third fully expanded leaf of the main stem. Ideally, when only a moderate number of accessions is analyzed, 3 plants of a bloc and 3 blocs per treatment are measured. When a large genotype set is under investigation, one plant per bloc is chosen, labeled and used for measurements at each time point.

For relative leaf water content analysis whole 3rd leaves of the main stem are sampled, covered with saran wrap and processed for RWC measurement. For leave water potential, leave discs are taken from the apical leaflet of the 3rd fully expanded leaf using a cork borer and processed in an osmometer.

The amount of destructive sampling for biochemical and genomic analysis depends of the measurements to be taken. Ideally, samples from the three central plants of a bloc are pooled. For leaf samples, the third leaf or an appropriate number of leaflets of the third leave is taken. For stolons, all stolons of a plant are pooled and from roots distal parts with about 2 mm diameter are sampled, rapidly cleaned from soil and rapidly rinsed with distilled water. For tuber sampling, slices of about 0.5 cm are cut from the middle of washed and unpeeled tuber in longitudinal and transversal direction. The plant material is dripped in liquid nitrogen immediately after harvest in order to "freeze" the condition of the plant, transported to the laboratory under liquid nitrogen and kept at -80ºC for genomics analysis or -20º for metabolome analysis. Samples for metabolite analysis are lyophilized and ground before extraction.

Data Evaluation

A phenotyping flow scheme of a typical droughty trial on potato is given in fig. below.

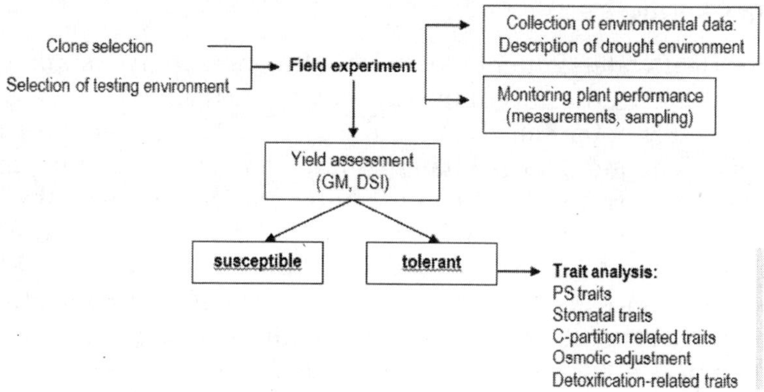

Environmental Parameters

Soil water content and vapor pressure deficit are key drought stress descriptors and are used to normalize quantitative plant parameters. Irradiation values serve to normalize porometry and photosynthesis measurements and temperature and humidity measurements result in vapor pressure deficit values that also serve to normalize phenotyping data.

Plant Development and Morphology

Drought influences plant development. Dependent of the genotype, timing and strength of drought, water stress might accelerate or delay flowering and tuberization, or slow down canopy growth and tuber bulking. It is important to measure differences in earliness between clones in order not to confound drought tolerance and drought escape. Recording morphological and phenological parameters assists in measuring the impact of drought on plant development. Leaf area is an important parameter for the normalization of photosynthesis or transpiration-related measurements. Determination of root size is reveals differences in access to soil water between clones.

Agronomical Analysis

Yield maintenance under water stress conditions is the principal measurement of drought tolerance. Yield data under drought and control conditions are used to determine the geometric mean (GM) of yield, which is a good predictor of genotype performance in stress and non-stress environments. GM is calculated as v(Yield under stress x Yield control). Further, yield data are used to calculate the drought susceptibility index, which is a measurement for the relative drought resistance of a genotype compared to the mean of the investigated accessions and is calculated as (1-Yieldunder stress/Yieldcontrol)/(1-mean under stress/meancontrol). Yield partition analysis reveals differences in carbon allocation to different plant tissues.

Plant Water Status

Relative leaf water content measures differences in hydratation status of leaf tissue and is an indicator for the capacity of the plant to maintain its water status. Leaf water potential reveals the water up-take capacity of the plant from a drying soil and also estimates osmotic adjustment of the plant tissue during drought stress. Stomata control transpiration and have the tendency to be closed under water stress in order to avoid water loss. However, closed stomata also inhibit CO_2 uptake and thus limit photosynthesis. Measuring stomatal conductance in connection with leaf area, relative leaf water content, and photosynthesis efficiency allows to identify stomatal and non-stomatal effects on photosynthesis efficiency under drought.

Transpiration efficiency is an indicator for water relations in the plant during soil drying. The sooner the amount of transpiration drops in a plant grown on a successively drying substrate, the lower is the capacity of the plant to extract soil water and thus the higher is its drought sensitivity.

Water use efficiency describes how much water a plant needs per unit of produced biomass or yield. Plants with high water use efficiency are desirable for drought tolerance breeding, however, there is a multitude of traits determining water use efficiency, such as gas exchange efficiency, carbon allocation, growth regulation and transpiration efficiency, which themselves are determined by many different traits.

Plant Health

Sensing spectral reflectance data of crop plants allows to detect crop stress in very early stages. Spectra of solar radiation reflected and absorbed by the crop canopy are measured to uncover information about a crop's health. Spectroradiometers monitor the continuous spectrum from 350nm to 1050nm. Recording data over treatments and genotypes and correlating these data to physiological and agronomical parameters allows to translate spectral information into of the crop's water or nitrogen status.

Photosynthesis

Plant productivity depends on the efficiency of photosynthesis. Drought lowers photosynthetic efficiency through increased photorespiration through lack of CO_2 due to stomatal closure. Determination of the status of photosystem II by measuring chlorophyll fluorescence and determination of CO_2 uptake by the plant are basic parameters to identify genotype-dependent differences in photosynthetic efficiency under drought.

Nitrogen Uptake

Nitrate reductase is a central enzyme of the nitrogen metabolism in the plant. Nitrate uptake and carbon assimilation are coordinated in the plant, but might be impaired under drought. A decrease in nitrate reductase activity in stressed plants indicates increased stress susceptibility of the plant. Nitrate reductase also catalyzes the formation of nitric oxide, an important messenger for abscissic acid biosynthesis and stomatal closure.

Hormone Status

Abscissic acid is a central hormone for drought stress signaling. It regulates drought responses such as stomatal closure, drought induced root growth and gene expression changes. Quantification of ABA over time is a good indicator for drought stress and can be related to plant drought responses.

Giberellins are hormones regulating plant development and are involved with tuberization in potato. Drought changes the gibberell composition in plant tissues, causing alterations in plant development, such as canopy size reduction or changes in tuberization timing and efficiency.

Genomics and Metabolomics

Determination of changes in metabolite content and gene expression provides a bulk of data giving insight into plant stress responses. Measurements are expensive and time consuming and thus are only practical for the investigation of a small genotype set, mostly for drought tolerance trait capture.

Metabolite analysis yields data on sugar and sugar alcohol accumulation and reveals changes of amino acid patterns in plants, which can be interpreted as osmotic adjustment under drought. Genotypic variation in metabolite concentrations yields important information on clone-specific differences in drought adaptation.

Gene expression changes monitored at different time points during drought pinpoint drought-induced functions over time. Microarray-based gene expression assessment is most informative, but also most expensive. It allows the monitoring of the expression of several thousands of genes, which can be attributed to functions and/or biochemical pathways. These data can be used to define candidate gene lists for drought tolerance.

A drought tolerance candidate gene list has been established for potato. This list contains a set of drought-inducible transcription factors driving expression of many drought response genes. An early and efficient up-regulation of these genes contributes to efficient drought adaptation of the plant. Similar conclusions can be drawn from the activation of specific signaling genes such as kinases or phosphatases.

Virtually all kinds of stresses cause accumulation of active oxygen species in the plant. Oxidation status is a crucial regulator of cellular functions, while excessive oxidative stress causes cell death. A bulk of genes involved in mitigating effects of oxidative stress is activated under drought. High expression of these genes is considered as a marker for stress tolerance.

5

Drought Hazard Assessment and Mapping

INTRODUCTION

The Organization of American States (OAS) as part of its Post-Georges Disaster Mitigation Project (PGDM) commissioned this drought hazard assessment and mapping study for Antigua/Barbuda.

One of the major objectives of the PGDM is the "development of national goals, objectives and actions to reduce the vulnerability of Antigua/Barbuda … to the effects of natural hazards."

TERMS OF REFERENCE

The Terms of Reference (TOR) for the study is attached as Appendix 1. There are two major component products of this assignment, (i) a drought hazard assessment report and (ii) island wide GIS data layers depicting drought hazard risk areas and drought hazard maps for Antigua/Barbuda.

METHODOLOGY

Data Collection

Data for map preparation and for drought assessment were collected using secondary sources, reports, maps, personal communication and limited field observations.

Mapping

Maps were first manually prepared and then digitized using GIS application, Arc View.

Definitions

For the purposes of the study and in agreement with relevant authorities, a definition of drought has been drafted, as indicated below. Other relevant definitions are also provided to facilitate both the drought hazard assessment and mapping work.

Drought

Drought is a recurrent feature of Antigua/Barbuda's climate. It occurs when there is an extended period of deficiency in precipitation (relative to what is considered normal), which is then insufficient to meet economic, social and environmental demands. Given their relatively small size, drought effects in Antigua and Barbuda are felt island-wide.

For reasons of analysis and in view of the need to determine appropriate responses to drought impacts and vulnerability, drought is defined as having three critical but inter-related components (as taken from the US National Drought Mitigation Center's definition) namely:

Meteorological drought

Meteorological drought is defined mainly by deficiencies in precipitation. Along with deficient rainfall, conditions during drought may be accompanied or aggravated by high temperatures, strong winds, low relative humidity, greater sunshine and less cloud cover. These conditions can be expected to bring increased evaporation and transpiration, reduced water infiltration into soils and a reduction in deep percolation and ground water recharge. The Antigua/Barbuda Meteorological Office (Met Office) has categorized drought from slight to extreme.

Table. Categories of Drought, as Defined by per cent of Average Yearly Rainfall of 40.98 inches (Antigua)

Category Drought	Rainfall (inches)	% of Ave. Annual Rainfall
Slight	38.93	95
Mild	36.88	90
Moderate	32.78	85
Moderate/ Severe	30.74	80
Severe	28.69	75
Extreme	24.50	70

Based on annual rainfall assigned to each category by the MET Office, some form of drought occurred in Antigua in 20 of the 40 years between 1960 and 2000. Rainfall records for the period were collected by the Met Office. This does not match public recollection and perception of drought. A more simplified definition of meteorological drought using the Met Office's Moderately Severe category, as the cut-off point would provide a definition more in keeping with public perception.

Meteorological drought for Antigua/Barbuda is defined using a precipitation level of 80 per cent or less of yearly average rainfall (or < 30.74 inches, Antigua and < 27.79 inches, Barbuda) for a drought year. A limitation in using this definition is that drought designation is mostly done on the basis of a calendar year and usually after the event occurs.

For the purpose of drought preparedness, it is suggested that precipitation deficiencies over a six (6) month period could be used to predict an imminent drought. For example, if the average monthly rainfall for the dry period of the year, January to April, (2.06 inches for Antigua) is experienced over a six (6) month period, it would in public perception signal that a drought was imminent. At the end of a six (6) month period in which total rainfall is < 12.36 inches, effects associated with meteorological drought would begin to emerge.

Agricultural drought

Agricultural drought occurs when plant water demands cannot be met due to soil water deficiency resulting from dryness brought on by meteorological or hydrological drought. In such cases plant water stress may be evidenced from reduced biomass and plant yield. One common indicator species is the "evergreen" ficus, which responds by first dropping most or all its leaves followed by the death of its stems.

Hydrological drought

Conditions associated with meteorological drought represent the earlier signs of drought followed by effects of agricultural drought because plants are highly dependent on stored soil water. Hydrological drought can be considered a third stage in the evolution of drought conditions evidenced by significant reduction in surface reservoirs, drying of dams and wetlands. In this phase of drought, livestock may be severely impacted and other sectors begin to feel the devastation.

Drought Hazard

A drought related hazard is an event in which a significant reduction of water brings about severe economic, social and environmental hardships to the population of Antigua/Barbuda.

Drought Vulnerability

Vulnerability to drought is defined as economic, social and environmental characteristics and practices of the country's population that make it susceptible to the effects of a drought. Vulnerability is reduced by the ability to effectively plan for, anticipate, cope with and recover from droughts.

Drought Risk

The potential adverse effects of droughts viewed in relation to their frequency and severity combined with the vulnerability of Antigua/Barbuda's population determines the risk to such events.

Use of the Term Drought

In this report, reference is made to the meteorological, hydrological or agricultural drought as appropriate. A simple working definition of drought could not be developed for this assessment because of varying perceptions of drought even for persons familiar with drought hazard assessment literature.

Broad-based discussions are needed between persons involved with drought mitigation planning and those involved with sectors affected by drought in order to build consensus and capacity to manage drought.

ASSESSMENT

METEOROLOGICAL CONDITIONS

Average annual rainfall for Antigua is 40.98 inches. Rainfall records for Antigua 1950 – 1984, as reported by the Island Resources Foundation (IRF), using APUA as its source, are summarized in Appendix 3. Records from 1960 – 2000 based on observations at Coolidge by the Met Office are given in Appendix 2.

Both are presented because of variations, which suggest that the data was obtained from two different stations. Attempts to determine the position of the rainfall station from which APUA records were taken were unsuccessful. Both records show significant variability in precipitation between years from 1950 to 2000. A rainfall map, showing "long average annual isohyetals" indicates important variability between sections of Antigua, with the volcanic hilly region in the south/southwest having up to 55 inches/year, while coastal areas in the southeast, east and northeast experiencing 35 inches/year.

Based on the average rainfall for coastal areas as cited by the isohyetals, precipitation >35inches/year (87.5 per cent of average annual rainfall) would not constitute drought conditions, or else such areas would be in a state of constant drought. IRF identified two periods between 1950 and 1990 as "periods of prolonged drought" for Antigua. These are 1964 to 1968, when the average rainfall was 31.53 inches, and 1983 to 1984, when the average rainfall was 27.75 inches. In the Met Office records, eight (8) years between 1960 and 2000 experienced rainfall below 30.74 inches and could therefore be considered drought years for Antigua.

Variation in monthly rainfall is another important consideration in drought management. The data show that precipitation is highest during the months of August to December (wet season) and lowest between January and April (dry season). Average wet season rainfall is 4.6 inches/month, while average dry season rainfall is 2.06 inches/month. Therefore, for four months of the year (January to April) the average rainfall of 2.06 inches is 60 per cent of average monthly rainfall for the 12 months of the year (3.41 inches) but the period is not long enough to be perceived by the population as drought. It is assumed however, that if dry season conditions were extended for six (6) months (12.36 inches of rainfall) this would, in public perception, constitute drought conditions. A critical constraint in the use of rainfall data is the distribution of rainfall stations. Stations are lacking in the drier areas of the southeast, east and north of Antigua, where shallow soils and the relative exposure to winds and marine influences tend to make such areas relatively more susceptible to the effects of drought. In past years there were 70-95 rain gauges located around Antigua (IRF, 1991). Most are no longer in use.

BARBUDA

The average annual rainfall for Barbuda between 1965 and 2000 was 34.74 inches. During this period, if < 27.79 inches/year of rain is used to define meteorological drought, then drought years were experienced in 1966, 1967, 1968, 1971, 1977, 1983, 1990, 1991, 1994 and 2000.

Temperature

Average temperatures are 29 degrees C (82.2 F) during the summer and 24 degrees C (75.2F) during the winter months. Extreme highs of 34 degrees C (93.2F) in August and 15 degrees C (59F) in January have been recorded (IRF). Despite their small size, both islands have topographic and landscape features (including urban development), localized wind patterns and soil characteristics that result in temperature variations over the island. Such temperature differences are not captured due to a shortage of temperature recording stations. Existing temperature stations record surface temperatures. Soil temperatures, which are critical to the performance of some plants and to agricultural crops, are not normally recorded.

Winds

Easterly trade winds form a critical component of Antigua/Barbuda's climate. Yearly average wind speeds between 1969 and 1995 recorded at the Coolidge Met Office, Antigua, was 12 knots. Highest average wind speed is recorded for the month of July (14.2 knots). Critically, the lowest average wind speeds are for the months of September, October, November, which are the three wettest months of the year.

Table. Rainfall, Windspeed and Relative Humidity, Driest and Wettest months of the Year

January	**2.34 Rainfall (inches)**	**12.8 Windspeed (knots)**	**81.0 Relative Humidity**
Dry months			
January	2.34	12.8	81.0
February	1.53	12.4	81.0
March	1.77	12.1	81.0
June	2.01	13.4	82.0

Ave.	1.91	12.7	81.25
Wet Months			
September	5.29	10.8	84.0
October	5.11	9.6	85.0
November	5.14	10.5	85.0
Ave.	5.18	10.3	84.6

Note: Average monthly rainfall data is based on records from 1960-2000. Average monthly windspeed data is based on records from 1969 – 1995 and average monthly relative humidity data based on records, 1960-1995.

Relatively speaking, winds are normally stronger in the dryer months when bare soils are exposed to its forces. Actual monthly windspeed data were not obtained so as to determine if this relationship holds true for extended dry periods (droughts). It does suggest that wind speed is one factor that makes soils more vulnerable to erosion during drought.

The records indicate that between 1991 and 1995 the prevailing wind direction was 090 degrees for eight months of the year – shifting slightly to 100 degrees for the months of April, June, and October and 110 degrees during the month of May.The drying effects of wind on soil is particularly critical for exposed east facing slopes of the southeast, east and northeast of Antigua and the Highlands on Barbuda's east coast.

The stronger winds of the drier months rapidly remove moisture from soils, affecting in particular plants with shallow root systems. The direct physical effects of the wind, combined with salt in the air contribute to the wilting of plants. Wilting also occurs from deficiencies in moisture. These localized effects have not been sufficiently studied although their understanding is critical to the assessment of areas susceptible to drought.

Wind data for Barbuda was not obtained but it is assumed that average wind speeds do not vary significantly from Antigua.

Relative Exposure of Slopes

The exposure of slopes to wind and sunlight affects vegetation and soils and rates of evaporation and transpiration. These influence localized climatic conditions, which may in turn affect conditions for grazing and growing crops. For example, light increases transpiration (evaporation of moisture from leaf surfaces) rates more than it does evaporation rates (from water bodies, soil). On the other hand wind increases evaporation rates more than transpiration rates (OAS, 1991).

East facing slopes, particularly those at the shoreline, therefore experience relatively higher evaporation and transpiration rates than western facing slopes, where the afternoon sun has a greater effect on transpiration rates than does exposure to winds.

The critically exposed east facing slopes near the coast in the Mamora/ Isaac Hill region of Antigua exemplifies the effects of wind and light, along with the effects of overgrazing by its barren landscape. Western facing slopes in the higher volcanic region of Antigua enjoy relatively better precipitation due to convective influences. In this case, warmer air rising over the land surface at the higher elevations cools and condenses releasing its moisture on slopes west and northwest at the Shekerley Mountains range. Here, the area of influence is relatively small and the elevation not high enough to cause the warming of descending air and subsequent withholding of moisture. As a result, the arid conditions experienced on the lee of mountains in Central America for example from orographic precipitation do not exist.

Relative Humidity

Average relative humidity for dry and wet months, based on MET Office records. In addition, the IRF (1991) reporting from Loveless (1960) and Atkins (1983) claims that the mean relative humidity for Antigua is in the low 80's in the morning and low 70's in the evening.

Relative humidity is said to be higher in Antigua than in Barbuda and some of the other islands. Heavy dews deposited at night on the island are believed to contribute to water balance, particularly in drier areas. The amount of water gained through this form of precipitation has not been measured.

Sunshine and Cloud Cover

Antigua/Barbuda experiences ample periods of sunshine throughout the course of a year including the wet season. The islands experience dense cloud cover (as different from scattered clouds) for much less than half of the year. Hence despite having average relative humidity higher than neighboring islands, Antigua is normally dry and temperatures relatively high. Because of the relationship of temperature to evapo-transpiration, rates for the latter are considered on par with other countries in the Leeward Islands.

EVAPORATION AND TRANSPIRATION

Evaporation

Evaporation rates for water bodies and soils in Antigua/Barbuda are not generally available. Evaporation and seepage rates were calculated for Potworks Reservoir (between 152 and 188 ha in reservoir area) during November 1995 to May 1996. The results, show that total evaporation and seepage for the seven (7) month period (1.59 million cu m) was over two times the amount of water supplied from the reservoir (691,091 cu. m). This is of critical significance to water management during drought.

Although evaporation rates and volumes were not separated, the results indicate the relative quantities of water than can be expected to be lost from surface water bodies including municipal and agricultural storage reservoirs and dams.

Table. Water Balance at Potworks Reservoir Nov 1995-May 1996

Ave. Rainfall (monthly)	46.428mm
Total Rainfall (7 months)	325mm
Reservoir Area	173.7 ha
Monthly ave. reservoir vol.	3.4 mil. cu. m
Total reservoir vol.	23.8 mil cu. m
Monthly rate evaporation/seepage	155mm
Monthly ave. evaporation/seepage	227,628 cu m
Total evaporation/seepage	1.59 mil. cu. m
Monthly ave. water supply	98,727 cu. m
Total water supply	691,091 cu. m

Comparable estimates for other surface water bodies were not available and no such studies have been done to estimate evaporation from soils. However, it is assumed that evaporation rates increase in areas where the soil has been exposed by overgrazing.

GENERAL ENVIRONMENTAL CONDITIONS

Geology

Antigua has a volcanic region (the hilly southern portion of the island) and a limestone region at the northern part of the island. A central plain, referred to as the third volcanic region, contains stratified volcanic detritus and agglomerates derived from the influences of the volcanic and limestone regions. Barbuda is of limestone formation. Geology influences soil type and vegetation. In the case of Antigua and Barbuda, volcanic formation affects and differentiates topography. Geology is therefore an underlying factor contributing to the assessment of drought impacts.

Slopes

Slopes in the volcanic region of Antigua are mainly between 11 and 20 degrees (steeper in some cases), while in the limestone and central plain region slopes are generally less than 10 degrees. In Barbuda, less than 1 per cent of the slopes are over 10 degrees and these are found in the Highlands. Slope classes are mapped on the drought hazard assessment slope data map for Antigua and on the slope data map for Barbuda.

Table. Slope Classes, Barbuda

Slope Class (degrees)	Area (acres)	%
0-2	8,396	22.87
2-5	18,514	50.41
5-10	9,794	26.67
10-20	20	0.05
Total	36,724	100.0

SOILS

Antigua

Hill et al did a detailed soil survey of Antigua and Barbuda for the College of Tropical Agriculture, UWI, Trinidad in 1966. They identified 33 soil types. Atkins (1983) reported by IRF (1991) grouped the 33 soil types for Antigua into four (4) general categories, namely:

- Deep, alluvial, colluvial soils found mainly in the narrow valleys of the volcanic region
- Deep, Kaolinitic clay found mainly in the central plain
- Shallow soils found in the volcanic region, and in the limestone region north of St. Johns (extending roughly from the airport in the east to Weatherills in the northwest)
- Complex of shallow and deep soils, found in the northeastern section of the limestone region

The soil data map for Antigua is based on these four (4) soil classes.

Deep alluvial soils of finer texture in the valleys of the volcanic region and the deep clay soils of the central plains and pockets of the limestone region have good moisture retention capacity. Plants in such areas will benefit if the depth to ground water provides a reliable supply of moisture without being waterlogged. With soil texture and position in relation to ground water conducive to moisture retention, and with soils less prone to erosion, such areas are less vulnerable to drought than areas of steep slopes and shallow soils.

The shallow soils of the volcanic region are neutral to slightly acidic and well drained. Moisture retention is relatively less than in deep alluvial or deep clay soils. However, moisture retention is improved in undisturbed areas where significant amounts of organic matter are deposited on the forest floor. Drought effects on soils in these areas are less severe than in disturbed open woodland areas with similar slope and shallowness of soil.

Soil loss from denuded landscape in areas with steep slopes and shallow soils is significant and are exacerbated by drought. Soil erosion is accompanied by a gradual loss of soil nutrients and plant cover recovery is slow or may not occur. This characteristic is obvious in areas that have experienced extensive clearing and grazing. Generally shallow soils are relatively more vulnerable to drought and the presence of shallow soils is therefore used as a criterion in mapping areas at risk to drought.

Soils in most of the central plain are well drained but there are areas of heavy clay and poor drainage. Heavy clays may hold moisture longer, which is helpful in dry periods, but restrict crop growth in normal conditions.

Soils in the limestone region are mostly alkaline and are light where there is an overlay of calcareous sandstones, heavy over calcareous grits and deep and well drained over calcareous marls (IRF, 1991). Well-drained soils lose moisture quickly so that plants in such areas reach wilting point relatively quickly in dry periods. As with the shallow soils of the volcanic region, the shallow soils found in some parts of the limestone region represent major management challenges for soil and moisture conservation where land has been cleared or intensively grazed.

Barbuda

Following the work of Hill et al (1966), soils of Barbuda were mapped by the OAS (1992). The OAS soil classification has been used in the preparation of the drought hazard soils data map for Barbuda. This classification places soils in six (6) series, namely:

- Codrington clay found main around Codrington and to its south and southeast extending to the coast, in flat areas (2-5 degrees). These soils are considered suitable for agriculture but are limited by shallowness, stoniness and in some cases salinity
- Barbuda clay loam found in the Highlands in 5-20 degree slopes. Lands are marginal or unsuitable to cultivation. A critical characteristic of these soils is moderate to severe erosion resulting in between 50 per cent to total top soil loss
- Blackmere clay loam found in approximately 1/3 of Barbuda, in areas east, southeast and northeast of Codrington, at the southeastern tip of the island, at Goat, Rabbit and Kid islands and at the narrow sand bar separating the west shoreline and Codrington Lagoon. Land in this series slopes <10 degrees and is characterized by varying limitations of shallowness, stoniness, salinity and compaction
- Beach sand found at beaches and overlaying the critical aquifer at Palmetto Point
- Mangrove, Swamp found in the mangroves of Codrington Lagoon and in the swamps at the southern section of the island
- Salina found fringing sections of the Codrington Lagoon and the major swamp in the south of the island

VEGETATION

Antigua

Horwith and Lindsay carried out a detailed vegetation classification for Antigua in 1990s but did not map their results. Attempts to obtain or review a copy of the report were not successful.

The Forestry Division has plans to undertake mapping of the island's vegetation. Vegetation zones have been altered in the last 20 years from clearing for residential developments and from extensive grazing by small ruminants. In the absence of a detailed up-to-date vegetation classification, mapping and digitization of vegetation was done using a combination of sources, along with field observations. Main sources were IRF (1991), who adopted earlier works and the OAS (1992) land use map.

Vegetation categories mapped (see drought hazard assessment vegetation data map for Antigua) are:

- Cactus scrub accounting for almost 2/3 of the land area and being the parts of the island in which most settlement and grazing activity occur. Cactus Scrub areas vary in percentage cover of Acacias, Lucaena, logwood and other woody species and in some cases have significant grass cover
- Dry woodland mainly found in sections of the east coast and pockets at the northern parts of the island. This vegetation category has been impacted severely by grazing and building and is being transformed to cactus scrub in some areas.
- Moist forest found mainly in the volcanic south and southwest, where higher elevations and steeper slopes limit access and human interference. While moist forest is the predominant vegetation type of this zone, areas of drier vegetation assemblages exist. Chances of maintaining the remaining moist forests in the volcanic region are good because land-use threats are less than in forested areas of the central plain and limestone regions. Failure to protect remaining forests will further deepen the island's vulnerability to drought. Since the abandonment of large-scale sugarcane cultivation in 1974 Antigua has not experienced any appreciable gain in new cactus scrub formation that would lead in succession to dry woodland and eventually moist forests where precipitation is adequate. Trends suggest that the percentage cover of Acacias and other wooded thorny species found in cactus scrub areas could further decline in the next decade and, as a result, further reduce total forest cover on Antigua.
- Mangrove woodland, Swamp found in coastal areas. The use of mangroves as wattle for fish-pots and for charcoal has declined in the last 20 years. Human impacts are mostly from cutting or filling for coastal development, while significant damage from hurricanes have been sustained. Grazing, which adversely affects other vegetation types, has had no direct physical impacts on mangroves and swamp vegetation.

Barbuda

Barbuda's vegetation has been mapped for this assignment (see drought hazard assessment vegetation data map for Barbuda) using the following categories:

- Dry woodland covering the Highlands, areas at the northern section of the island and at the west coast sand bar separating the sea from the Codrington Lagoon.

- Cactus scrub covering most of the rest of the island.
- Coconut plantation in the Palmetto Point and near Coco Point.
- Mangrove fringing areas of the Codrington Lagoon.
- Salina vegetation fringing the Codrington Lagoon and Goat Island Flush.
- Swamp vegetation.

SOIL WATER DEFICIENCIES

OAS (1992) provides estimates of evapo-transpiration and irrigation requirements for the Green Castle and Coolidge regions of Antigua and the Codrington region of Barbuda. Based on the results irrigation would be required for the months of February and March, June and July for the three areas, to counter moisture deficits resulting from higher evapo-transpiration rates.

While actual evapo-transpiration rates are not available to most farms, crop and livestock farmers and ornamental gardeners are very much aware of the rate at which plants reach their wilting point in shallow soils where little or no water is available during dry spells.

Deficiencies in soil water is readily discernable by a deteriorated "brown" landscape, particularly obvious in areas of extensive grass cover over shallow well drained soils. Even when water is available through irrigation, the rate of soil water loss increases where water conservation measures such as mulching or the use of windbreaks are not used.

Where soil water loss leads to reduced biomass in grass or plant species that are normally readily available to small ruminants, goats in particular browse beyond their normal range feeding on crops and ornamental plants. In such cases adverse environmental conditions lead to economic or financial loss to affected persons.

WATER RESOURCES

Antigua

Watershed boundaries for Antigua were mapped using IRF (1991), who quoted McMillan (1985) and OAS (1992). According to IRF, eighty-six (86) watersheds were initially defined by Halcrow (1977) and these were reduced to thirteen (13) by McMillan.

These same boundaries have been used by the OAS (1992) and for this assessment (see drought hazard assessment watershed data map for Antigua). Each watershed has a defined drainage system and can be considered a hydrological unit of utmost relevance to drought management. Watershed 1, which drains to Nonsuch Bay on the east coast and Watershed 2, which drains to Hanson's Bay on the west coast are two of the most significant watersheds for the storage of surface water for domestic and agricultural use. Upland stream-flow at watersheds draining the southwest volcanic region is less affected by land use, in particular grazing, than most other watersheds. Stream-flow is not permanent for all watercourses in Antigua, although Cooks Creek that flows through Creekside maintains water for most of the year.

Storage capacities for all municipal and agricultural reservoirs are given on the Watershed map for Antigua and summarized.

Barbuda

Ten watersheds have been identified for Barbuda (see drought hazard assessment watershed data map for Barbuda), although boundary definitions are not finite due to its unique topographical features. Watershed 1 includes the Palmetto aquifer – considered at one time to be the best quality aquifer on the island but has been contaminated by sand mining and grazing. Drainage from all other watersheds originates in the Highlands.

A number of wells are shown on the watershed data map for Barbuda. In addition to these are numerous wells dug and used by individual households not linked to municipal water supply. A number of sinkholes exist in the Highlands area. These include sinkholes at Brian Cave and Dark Cave, both with depth to the water table (J. Mussington, pers. comm.).

The above suggests that the fresh ground water reserves are extensive in Barbuda but poorly managed. There are no records on the number of private household wells and pumping is unregulated. Management of fresh water resources in Barbuda would seem to be a critical issue requiring immediate attention. An initial effort should be the mapping of all privately used wells. In the longterm two initiatives should be considered.

One is a comprehensive investigation of fresh water resources; there is no indication that this has ever been done. The other is an appropriate regulatory regime seeking to manage ground water against contamination from land uses and over-extraction by private households.

LAND USE AND LAND MANAGEMENT PRACTICES

CROP AND VEGETABLE FARMING

Antigua

Crop and vegetable farming occur in pockets throughout the island. Some of these pockets are located in the narrow valleys with deep alluvial soils of the volcanic southwest. Others appear just as often in the central plains and limestone regions where the topsoil is less thick.

Apart from valley topography there are no readily discernable spatial patterns to cultivation and there are no farm belts. Precipitation, moisture retention and pH often determine the choice of crop so that most of the bananas grown on farms are planted in the valley below Fig Tree Hill, an area receiving relatively good rainfall and protection from wind.

Most of the pineapples on the island are planted in the valleys of the volcanic region where soils tend to be neutral or slightly acidic in contrast to the more alkaline soils of the limestone region.

The moisture-loving banana is planted in numerous household gardens. Mango trees seem to thrive better in areas of higher rainfall and are quite obvious along the Fig Tree Drive. One thing noticeable about the location of farms is the relative distance of some in the volcanic region from the main road. In some of these cases infrastructure costs for water connection for irrigation may be prohibitive. Farms in the central plain and limestone regions are generally better positioned in relation to the main road and water mains.

TYPE OF CROPS

The drought hazard assessment land use data map prepared in this assignment grouped crop and vegetable farming and tree crops under the land use designation, agriculture. Under the OAS resources assessment project, cultivation uses were detailed by type and watershed. The approximate acreage for each use was also given. The types of uses are provided below, along with an indication of where crops are mostly grown. Reference made to the main locations of each category is drawn from OAS (1992) with some slight modifications. It is assumed that the acreage under the various crops has since changed but that the relative distribution of crops among watersheds has not changed significantly:

- Sugar cane grown mostly in watersheds 54-62 and 63-66. For good yield, sugarcane requires abundant moisture and deep adequately drained soils. Growing it in these relatively dryer watersheds of the limestone region in areas with shallow soils increases vulnerability to drought.
- Cotton grown in watersheds 1, 54-62 and 67-77.
- Coconut grown mainly in watersheds 2, 4-11, 12-20, 63-66 and 78-84. Plants are mostly found in sandy soils but coconuts grow well in other soils also, as long as there is an adequate supply of ground water with the water table being 150 – 180 cm (60 – 72 inches). Conditions are favorable for coconuts in the areas grown, so that it is perhaps the least vulnerable to drought among perennial crops.
- Fruit trees mainly mangoes, 63 per cent of acreage in watersheds 4-11 and 12-20. Higher rainfall and pockets of deep soil in these volcanic area watersheds are advantageous to perennial tree crops, which require soils with good moisture retention and depth to allow tap root development.
- Pineapple grown mainly in watershed 12-20. Acidic and well-

drained soils of this watershed are suitable for pineapples, which could be grown in other areas where similar conditions exist. Pineapples do not require soils with high moisture retention and is therefore less vulnerable than many crops to dry conditions.

- Banana grown mainly in watershed 12-20 but can be found in household gardens around the country. The plant favors the deep fertile soils with adequate moisture of the valleys in the volcanic region. Bananas are over 90 per cent water but moisture deficiencies during droughts retard plant growth and fruit development.
- Food crops including yams, sweet potato and maize, grown in all watersheds except 3, 78-84 and 85-86. These crops have varying soil and moisture requirements. Vulnerability to drought depends on where they are planted;
- Vegetable crops including eggplant, tomato, sweet pepper, beans and okra, grown in all watersheds. Vulnerability to drought is relative to location, soil type, moisture and other factors.

Barbuda

Food crops are grown in pockets on soils with limitations such as shallowness, stoniness, salinity or low rainfall. These limitations suggest high vulnerability to drought. Pockets of food crops are shown on the drought hazard land use data map for Barbuda. Food is also grown in Palmetto Point but the area could not be mapped because of insufficient data. It should be understood that shifting cultivation due to rainfall deficiencies and shallow soils is a feature of farming on the island.

Shifting agriculture occurs in the low and highlands, where the practice is to fence an area, then clear and burn the vegetation before planting (J. Mussington, pers. comm.). Areas are abandoned when conditions are no longer suited to productive farming. Coconuts are grown in sandy soils where the water table provides a reliable supply of moisture. Vulnerability to drought is low.

Farming Practices

Crops unsuitable to some areas are in many cases planted because land tenure reduces options and choices available to farmers. However, where crops match land capability, farming practices in which little attention is paid to soil and moisture conservation increase vulnerability to drought.

LIVESTOCK FARMING

Livestock Population

The livestock population in Antigua and Barbuda in 1984. Figures obtained from the Veterinary Division on current livestock for Antigua indicate a significant increase in the goat and sheep populations. The goat population for Antigua in 2001 is four (4) times its size in 1984. Today's sheep population in Antigua is also almost four (4) times the number for 1984. Figures provided by the Livestock Division show significant declines in the sheep, goats and cattle populations in Barbuda to 146, 252 and 29 respectively in 2001. This suggests that the impact of grazing on the landscape has been less in recent years.

Table. Livestock Population in Antigua/Barbuda, 1984

Livestock	Antigua	Barbuda
Cattle	9,992	1,072
Sheep	5,619	473
Goats	9,319	229
Pigs	2,425	12
Chickens	19, 554	529
Horses	179	184
Donkeys	713	215
Total	47,801	2,714

Table. Cattle, Sheep and Goat Populations Antigua, 2001

Livestock	Population
Cattle	15,364
Sheep	21,294
Goats	37,995

LIVESTOCK DISTRIBUTION

Antigua

Grazing areas have been mapped in this assessment. Grazing designations on the map follow those used by the OAS in its Natural Resources Assessment Project report, 1992 and are as summarized:

- Improved pasture: former cultivated lands, now managed in most cases (sometimes fenced) to control grazing. Antigua Hay grass and Seymour grass (Bothriocloa pertusa) are the dominant forage grasses. Acacia and other thorny species are maintained at less than 10 per cent cover. Major areas of improved pasture can be found along the Factory Road, St. Phillips and south and southwest of Creekside. Improved pastures occur on flat to gently sloping lands.
- Rough Grazing: former sugar lands, consisting mainly of Antigua Hay and Seymour grass and with Acacia cover in excess of 20 per cent (OAS, 1992). This form of grazing is the most extensive in the island's central plains, occurring on deep as well as shallow soils. Rough grazing is also prevalent the northern limestone area of the island. Grazing lands are not fenced

- Mixed Scrub and Rough Grazing: Abandoned sugar lands at the stage of vegetation succession where Acacias and other thorny scrub are dominant. This type of grazing occurs mostly in the central plains and limestone region. It also occurs extensively in the volcanic area between English Harbour, Isaac Hill and Mamora Bay.
- Woodland Grazing. This category is not mapped or described fully in the OAS report. OAS does make reference to small ruminants grazing on slopes with scant bush cover in volcanic areas. The extent of this form of grazing is however more significant than implied.

Large populations of goats (mainly) and sheep graze extensively in the Shirley Heights/Block House area and on wooded slopes of varying thickness in vegetation cover north and west of Falmouth. Grazing by goats in rough pastures, mixed scrub or wooded areas alters and degrade vegetation zones, accelerates soil and nutrient loss and renders areas susceptible to impacts from drought.

Barbuda

Small pockets of improved pasture are mapped for Barbuda on the drought hazard land use data map. This map also indicates a fairly extensive area of the lowlands designated mixed scrub and rough grazing. Most of the grazing by goats, sheep and cattle occur in this area but these animals also range in the highlands. In addition, deer, and feral pigs, donkeys and horses also graze the highlands.

LIVESTOCK MANAGEMENT PRACTICES

Antigua

Improved grazing occurs on lands owned, leased or rented by farmers, who shoulder capital and recurrent expenses for managed livestock operations. Cattle are common to such operations but many also manage sheep. Tethered cattle and sheep represent another form of management without fences done by landless or land owning farmers. There are however, thousands of goats that are owned by landless farmers who exercise little control over where the animals graze. In fact, goats grazing some of the volcanic slopes are in some cases considered "wild." Goat and sheep farmers tend to ignore damages caused by their livestock to cultivated crops and gardens. They also seem to be oblivious to the environmental damages caused by overgrazing. Existing penalties are not effective enough to halt inimical grazing practices. The tradition of impounding livestock caught damaging crops and gardens no longer exist because most, if not all, the pounds have been abandoned.

Barbuda

Untethered livestock is less of a problem to farms and gardens in Barbuda than in Antigua. Also, the decline in livestock populations since 1984 would have reduced the impacts on the landscape and habitats normally associated with overgrazing. Nevertheless, grazing remains largely uncontrolled. Its impacts on the vegetation of the highlands, where shallow soils is already a constraint to plant growth, is an aspect of landscape and habitat vulnerability to drought that should be of major concern.

MARKETS AND PRICES

One of the reasons for the significant increase in the sheep and goat populations between 1984 and 2001 is the existence of a strong local market. The price for mutton and lamb in Antigua is EC$7 per lb retail and live weight animals fetch an average of EC$3 per lb.

The demand for lamb and mutton, along with low overheads for managing un-tethered small ruminants and a weak policy to deter harmful grazing practices, provide conditions fertile to the continued growth of the sheep and goat populations in Antigua.

Settlements and Communities

Population

The population of Antigua decreased from 64,129 in 1970 to 58,114 in 1991, while that of Barbuda increased to 1,241 from 1,071. Population declined in all parishes but major losses occurred in St. Phillips, which lost more than half its 1970 population and accounted for 36 per cent of total population decline in Antigua. Other parishes with significant declines were St. Mary's 25 per cent and St. Paul's, 16 per cent of total population loss between 1970 and 1991. A breakdown of the 1991 population and population density by parish. Population density for Antigua is 538 persons per sq. mile but in St. John's City, the density is 7,419 persons per sq. mile. The high population density for St. John's City makes it relatively more vulnerable to drought than other settlements with respect to water demand and supply.

Many households, particularly in areas such as the Point, Gray's Green and Ovals do not have much space for above ground water storage and therefore must rely almost completely on piped water without the benefit of storage, a critical requirement when water must be rationed.

It is interesting that while the population decreased, the number of households increased from 15,405 for both Antigua and Barbuda in 1970 to 18,808 in 1991. This correlates with the visible growth in new housing within the 20-year period. Data on new housing, which should be available from the DCA, is not reliable because a significant number of new houses were constructed without planning permission in the period.

Given the decline in national population figures, population growth per se does not adversely affect the country's ability to deal with drought.

However, continued entrance of migrant workers to Antigua can increase vulnerability in areas of St. John's poorly equipped with water storage infrastructure. These are the areas that new migrant workers are attracted to because of more affordable rents and a convenient location to downtown St. John's.

In addition, the growth of new households is being accompanied by per capita increases in water use as the material well being of residents improves.

Settlement Distribution

Settlements are distributed island wide but most of Antigua's population lives in the central plains and limestone region. With the exception of John Hughes and Sawcolts, traditional settlements in the volcanic region are located mainly in the valleys or foot of slopes. Traditionally the village developed as a core entity surrounded by farmlands. Crop and vegetable farming were for most persons the primary agricultural activity linked geographically and socio-economically to households.

Subsistence livestock farming has now replaced crops and vegetables as the farming activity in which household members can be involved while working full time at other jobs. This is partly responsible for the increase of goat and sheep populations and the devastation that leaves the landscape more prone to drought impacts.

Residential Development Trends

Residential trends are changing the relationship between village and farm in Antigua. The same is not yet true for Barbuda.

In Antigua, government's policy of providing liberally sized housing lots (1/2 to 1 acre lots) often below market value is one of the factors that precipitated a trend towards lower density residential development and the growing urbanization of formerly cultivated lands. Another factor is the increase in vehicular ownership allowing mobility to the population and less dependence on public transportation. Infrastructure to serve low-density residential growth is more costly than for compact settlements and has been subsidized by government to a large degree. The implications have not been studied neither with respect to the financial burden on government or the long-term effects on land and environmental management. With regard to this assessment, the cost of infrastructure for water distribution is a factor that impacts on the country's capacity to manage water needs in dry periods and more so in drought conditions.

Hotels and Tourist Zones

Distribution

Tourism is Antigua's main foreign exchange earning industry. Hotels constitute the major tourism sub-sector but cruise and yachting tourism also contribute significantly to tourism earnings and jobs.

Beach hotels and other vacation accommodation constitute the major component of the tourism plant's land-based accommodation. Major tourism zones are Antigua's north coast, Five Islands' peninsula (including Deep Bay and Hawksbill Bay, Jolly Beach/Jolly Harbour), Long Bay Coastline and Falmouth Harbour/English Harbour.

Table. Tourism Accommodation in Antigua and Barbuda

Accommodation	# Properties	# Rooms
Antigua		
Hotels	40	2644
Guest Houses	8	59
Apts/Cottages/Villas 42	291	
Sub total	90	2994
Barbuda	4	111
Total	94	3105

Table. Distribution of Hotel Properties by Watershed

Watershed Antigua	# Properties	# Rooms
• 67-77 (Northcoast)	31	1104
• 78-84 (Five Islands Peninsula)	6	525
• 4-11 (Jolly Harbour/ Jolly Beach)	6	704
• 27-46 (Falmouth/English Harbour/St. James	8	292
• 54-62 (Longbay Coastline)	4	220
Other	45	149
Sub Total	90	2994
Barbuda	4	111
Total	94	3105

Since the effective management of land uses within the hydrological unit of the watershed helps to mitigate vulnerability to droughts, the distribution of hotel properties and rooms by watersheds.

Ninety five percent (95 per cent) of tourist accommodation rooms fall in five (5) watersheds, two of which are in the volcanic region, two in the limestone region (except for properties in St. John's) and one in the central plain. In general hotels do not affect upland run-off or erosion, since most are located at the coastline. Rather, hotels can be affected by upland run-off and erosion problems resulting from other forms of land use. Such problems may include pollution and turbidity impacts on bathing areas. Overgrazing can cause both. It also degrades the landscape, which adversely impacts the quality of the visitor experience, particularly during droughts. For example, Watershed 27-46 (Falmouth/English Harbour/St. James) with its 292 rooms has one of the largest population of goats and sheep on the island and perhaps the highest density of grazing and browsing small ruminants. Portions of the landscape seen from the scenic drive to historic Lookout (Shirley's Heights) and Block House have been devastated. Drought now has a more telling effect on the area. On the other hand, demand for water by hotels, yachts and cruise ships has grown. Per capita water use by stay-over hotel visitors is significantly more than resident demand; up to 300 gals/ day/person in one study against an average of <100 gals/day for residents.

INFRASTRUCTURE

WELLS

Antigua

Ground water in Antigua is drawn from about 50 active wells, located in several well fields. Most of the municipal and agricultural water in Antigua are stored in six (6) watersheds. According to the Ministry of Planning (1997), these watersheds also account for 80 per cent of ground water yield from wells.

Table. Major Well Fields of Antigua

Well Field	# wells	Location	Capacity (igpd)	Target Area
Bendals	13	Bendals	n/a	Ben-dals Valley
The Valley	20	BlubberC hris-tian, Roses, Oragne Valleys	n/a	n/a
Cades Bay/ Clare-mount	5	Cades Bay/Clare -mount	30,000	Old Road, Urlings John-sons Point, Crabb Hill
Follies	5	John Hughes/ Sweetes	40,000	N/a
Bristol Springs/ Collings	7	East of Collings Reser-voir	150,000	Willi-kies New-field St. Phil-lips Mill Reef Bethes-da

Barbuda

Well water is supplied by APUA to households in the central section of Codrington. Information on the number of households connected to mains supply or on the volume of water supplied was not obtained.

Numerous private wells exist and the water is used for bathing, flushing toilets and for watering gardens. The water quality is not monitored. Water for drinking and cooking is taken mainly from cisterns, even for households with mains supply.

DESALT AND OTHER WATER TREATMENT PLANTS

Two desalination plants, both located at Crabbs Peninsula, produce water for the national supply network. A flash distillation plant owned by government produces 2.8 million gpd and a reverse osmosis plant, owned by a private company, produces 800,000 gpd.

The government owned desalination plant was commissioned and built after the 1983-84 drought when Potworks Dam, which provided over 50 per cent of municipal water, went dry.

Desalinated water currently meets 62 per cent of Antigua's municipal water needs. This helps to cushion the country from water related shocks associated with droughts but production costs are much higher than other sources.

Production costs cited by the Ministry of Planning, based on information from the Antigua Public Utilities Authority (APUA), are:

- Ground water: US$2.50 per cu. meter
- Surface water: US$3.00 per cu. meter
- Desalinated water: US$4.70 per cu. meter

Water production cost by APUA is far in excess of water revenues. This is partly the result of the high cost of desalinated water but government's policy to keep water rates in check is part of the reason.

Predictions by the Ministry of Planning (1997) shows a continued negative gap between water revenues and production costs up to the year 2020, averaging over US$8 million per year. Water subsidies cannot be sustained forever, so that investment to increase ground and surface water supplies is being considered. This would help to reduce the negative gap between revenues and production costs but would not be the full solution.

The uncertainties over future climate change, variable precipitation and the continued growth in the demand for water means a continued reliance on desalination during average rainfall and more so during periods of drought.

Private Reverse Osmosis Plants

Several hotels now have reverse osmosis (RO) plants. Some were installed as a result of water shortages experienced in past droughts. Curtain bluff has a 50,000 gpd RO plant. An RO plant located on the Jolly Harbour property but operated by an independent contractor has a design capacity for 120,000 gpd but produces and sells an average of 70,000 gpd to the Jolly Harbour Resort complex. The supplier is able to offer water at lower rates than APUA, suggesting that APUA should be able to produce desalination water at lower than its current costs. Feed water to supply RO plants is taken from the sea in most cases. At Jolly Harbour 20 per cent of the feed water (35,000 gals per day) is drawn from a brackish well on the property. The rest comes from the sea. If the yield of the well drops during periods of hydrological drought a higher percentage of feed water is drawn from the sea at the marina basin. RO plants therefore reduce the vulnerability of hotels to meteorological and hydrological droughts.

WASTEWATER TREATMENT PLANTS

Few hotels on the island have water treatment plants that allow the reuse of treated wastewater. Jolly Harbour has a plant with design capacity to treat 500,000 gpd. The current wastewater load at Jolly Beach Hotel and Jolly Harbour Villas – the main users of the plant – is less than 120,000 gpd at high occupancy and less at lower occupancy. The treated wastewater is used to irrigate an 18-hole golf course. This significantly reduces the reliance on potable water for irrigation and brings the cost of water down.

IRRIGATION SYSTEMS

Agriculture

The acreage of cultivated land that currently has access to irrigation water could not be determined. Water sold by APUA at the agricultural rate has been averaging 198,000 cu. m per year, based on figures provided by the Ministry of Planning. The high cost of municipal water used for agriculture acts as a deterrent to irrigation.

Hotels and Golf Courses

Most hotels use some form of irrigation system for landscaping. Smaller properties use either hand held hoses or moveable sprinkler systems. Larger properties may have pop-up sprinkler systems or in some cases micro-irrigation systems (drip) using simple "soaker" hoses.Unfortunately, there is no available data on the percentage of properties using one system or the next. Environmental audits done for a number of hotels in the region indicate significant wastage of water from inappropriate water use, including poor irrigation practices. Where selected plants do not match soil and microclimatic conditions the tendency is to compensate by excessive watering. This happens in Antigua, although the extent is not known, and will continue until properties adopt more conscious water conservation methods.

Water requirements for golf courses can be in excess of 150,000 gallons per day for 18-hole courses, depending on the quality sought for putting greens and fairways. The course at Jolly Harbour uses between 90,000 and 125,000 gallons/day. The actual amount depends on rainfall. Fortunately, most of this is met by treated wastewater, which is adequate as long as occupancy levels remain above 70 per cent. Except for periods when low occupancy levels fail to generate enough wastewater volume, the golf course does not make any significant demand on national water supply.

WATER DISTRIBUTION

Variable pressure in water distribution creates problems for consumers. High pressure in some areas causes frequent leaks on the mains and consumer distribution lines. Low pressure in other areas results in inadequate flows and inconvenience to users.

Roof Catchment and Storage Cisterns

Roof catchment and storage in concrete cisterns cistern, metal or plastic tanks or metal drums is a traditional and accepted practice, particularly for homes. It is now a planning requirement for new developments to make adequate provision for water storage on property. The standards for residences and hotels now being used by the Development Control Authority (DCA) are: Residences: 10 gallons per sq. ft. of roof area. Hotels: 10 gallons per sq. ft. of roof area. Where hotels use reverse osmosis (RO) plants to augment water supply, the minimum standards for roof catchment/storage are less rigidly applied.

Water Supply and Demand

Water supply for Antigua and Barbuda averages 4.5 million cubic m per year from sources: According to the Ministry of Planning 42 per cent of total water supply was unaccounted for, 31 per cent was used by domestic consumers, 17.7 per cent by commercial and hotel users and 4.4 per cent by agricultural users. The Ministry of Planning in its 1997 climate change water resource sector study predicts that total water demand will be:

Year 2000: 4.2 million cu meters/yr

2010: 5.8 million cu meters/yr.

Table. Water Supply, Antigua/Barbuda, 1996

Source	Volume	per cent Total (Million cu. m/year)
Ground water (APUA)	0.6	13
Surface water (APUA)	0.9	20
Private roof catchment (Estimated)	0.1	2
Private dams (Estimated)	0.1	2
Desalination (APUA)	2.8	63
Total	4.5	100

It is presumed that the available supply would increase by reducing the amount of water that is lost from the system. Assumptions made by the

Ministry of Planning are that the population would grow to 72,300 by 2000 and 81,000 by 2010, and that per capita domestic water consumption would remain at 0.63 cu meter/day through to 2010.

These projections require careful monitoring because there is a degree of uncertainty associated with water demands – one such reason being the difficulty in predicting growth in the tourism sector. Water demand management therefore is major issue for drought mitigation for two reasons:

- Insufficient information on unaccounted for water.
- High production costs and hence high consumption costs, which are believed to mask true demand from the agricultural sector.

IMPACT AND VULNERABILITY OF DROUGHT

This section of the report reviews the impacts of past droughts on the islands and the response to the adverse effects experienced.

Description of Drought Events

Based on the meteorological definition for drought, a total of eight (8) droughts occurred in Antigua and ten (10) in Barbuda between 1960 and 2000.

Table. Drought Years in Antigua/ Barbuda 1960 – 2000

Year Rainfall (Inches)	Antigua	Barbuda
1966	27.52	12.89
1967	29.14	22.87
1968	25.09	27.79
1971	--	26.37
1973	25.98	--
1977	--	25.65
1983	22.31	20.67*
1990	--	24.17
1991	30.49	23.64
1994	30.49	25.04
1997	26.46	--
2000	n/a	23.30

* Rainfall for December not included

Between 1966 and 1968 there were three consecutive years of rainfall below 30 inches in Antigua. However, the 1983 drought, which extended into 1984, was the one that most people in Antigua remember.

It lasted from January 1983 to August 1984. In this 18-month period, monthly rainfall exceeded three inches twice, in August of 1983 and March 1984. Total rainfall for 1983 was 22.31 inches and for the first eight (8) months of 1984, 16.48 inches. The drought was ended by 8.54 inches of rain in September 1984. Between September and December 1984, 29.38 inches of rain were recorded. Because of its severity and impacts, this drought could be ranked as the "drought of record" for Antigua, meaning the drought remembered as having the greatest impact on the island. This drought and its impacts are described in the following pages. The impacts described are mainly for Antigua.

Rainfall records indicate that Barbuda experienced its severest drought in 1966 during the period, when rainfall for the year was 12.89 inches. Unfortunately, a review of the impacts of this drought was not possible during this assessment.

DROUGHT IMPACTS

WATERSHED AND HABITAT DEGRADATION

Watershed Impacts

Antigua received only 22 inches of rain in 1983, Barbuda less up to November of that year. By the time the drought entered its second year, marked changes in watershed and vegetation conditions were obvious.

Drought impacts were particularly severe in watersheds, such as 27-46, 47-53, 54-62 (see watersheds data map, Antigua) where plants with shallow root systems on shallow soils were among the first to be affected by deficiencies in soil water. For undeveloped areas of such watersheds seeking to re-establish an eco-system integrity after years of cultivation and charcoal burning significant losses in biomass productivity were experienced. This occurred even in thorny woodland vegetation associations, where acacias (Acacia spp.), logwood (Haematoxylum campechianum) and wild tamarind (Leucaena glauca) species are well adopted to shallow soils with low fertility and moisture deficient conditions.

No studies were made of soil conditions during the drought. It is assumed that moisture deficiencies eventually led to reduced soil fertility. The combined effect of water deficiency on shallow as well as deep root plants was progressive decline in the plant conditions.

Leaf and stem loss in herbaceous and woody plants were followed by plant mortality in certain species. As plants wilted and died, canopy cover decreased in wooded areas and the landscape generally became visibly brown, dusty and harsh looking. In Antigua/Barbuda, prolonged periods in which the landscape is visibly stressed are linked to hard-times; a disposition that can be traced to a time when rain fed agriculture provided the major source of income and livelihood for rural families.

The reduction and, in some cases, disappearance of plant cover exposed soils to the erosive force of winds and water when it eventually rained. Soil lost during the drought and the subsequent reduction in soil fertility, interrupted the process of ecosystem evolution occurring in watershed habitats prior to the drought event.

Habitat Degradation

In Antigua/Barbuda it is difficult to distinguish between habitats for wildlife and rangelands for wandering and uncontrolled livestock. For example, because parts of the landscape of the area delineated as Nelson's Dockyard National Park is steep and not readily accessible for agriculture and logging for charcoal they provide reasonable seclusion for birds, amphibians (frogs) reptiles (lizards) and mammals to reside. The Highlands of Barbuda is a critical habitat for the Fallow Deer (Dama dama) and other mammals. However, these areas also serve as rangelands for extensive grazing of goats and sheep. Livestock population densities in these areas are in excess of what is considered sustainable for like ecosystems. Before the drought, habitats were already susceptible because of overgrazing and thus suffered doubly with the extended periods of below normal precipitation. Habitat conditions then deteriorated for both wildlife and livestock in critical watersheds. One constraint in assessing habitat impacts from grazing is the absence of detailed listings of plant species preferred by goats. Unlike cattle and sheep, goats browse (rather than graze) normally on various herbaceous plants and woody shrubs along with ornamental plants if they gain access to household gardens.

Since biomass production of grass used by livestock in Antigua declines faster relative to many species used by goats, grazing livestock are likely to reach the critical threshold of nutritional stress earlier than goats. Thus, of the 47,801 livestock said to be in Antigua and 2,714 in Barbuda in 1984, goats are likely to have been the ones with lowest mortality from the drought.

It is not known if these figures reflected mortality associated with the drought or if most of the livestock mortality occurred after the field surveys for the census were done.

ECONOMIC IMPACTS

Agriculture and Food Crops

The 1984 agricultural census of Antigua and Barbuda identified 6,000 farmers (70 per cent of which were part time) in the country, 4,658 farms, and a total acreage farmed of 6,225 acres. (IRF, 1991). Landless farmers operated about 50 per cent of the farms and 40 per cent of the farms were under two acres. There were 66 farms in Antigua that were larger than 10 acres.

It is assumed that these figures represent the existing situation at the start of the drought. Also of importance to the understanding of drought impact is the breakdown between commercial and non-commercial production, namely:

- Thirty percent of production was done for home use only
- Thirty-three percent for home use mainly
- Thirty-seven percent for commercial use, mainly for domestic consumption

A range of food crops and fruits comprise most of the agricultural production. At the time of the drought cultivation of crops for export, such as sugar cane and cotton had been discontinued. Data from the census indicate significant declines in produce after 1984 for some crops, such as bananas, pineapples (448,000 lbs in 1984, 278,000 lbs in 1988), sweet potatoes (1,076,000 lbs in 1984, 482,000 lbs in 1988). On the other hand, crops such as mangoes (401,000 lbs in 1984, 1, 322,000 lbs in 1988), beets, grapefruits, limes and eggplants showed marked increases after 1984.

The census data does not correspond with anecdotal accounts in all cases. For example, one would expect that banana being a crop heavily dependent on water, would have suffered declines in yield between 1983 and 1984. Most farms in Antigua/Barbuda were at the time of the drought (and still are) rain dependent. The number of irrigated farms was not known for the time. As a result, widespread crop failure was reported as a consequence of the drought. Even for farms with access to potable water, rationing of supplies meant that priority for scarce water was given to hotels and households.

Livestock

The number of livestock that died during the drought was not determined. Residents however remember images of cattle and sheep lying along the wayside. The rains that ended the drought also took the lives of several cows around the Potworks Dam area. It seemed that some of the cattle were too weak from malnutrition to avoid the floodwaters and were washed along the watercourses. The drought affected the health, weight (hence marketability) and productivity of livestock. Hot weather caused heat stress, which is said to reduce fertility through sperm damage. Hot weather also affects libido in male animals, decreases milk production in cows and kills chickens. Animals also suffered from dehydration caused by shortages of water, when dams dried and water flowing through the pipes was not enough to meet the growing demands.

In some cases, livestock farmers killed their animals to avoid death from the effects of the drought. In other cases, animals that died from drought effects were butchered and sold without the knowledge of consumers, causing great alarm when this was discovered.

The economic impacts could therefore be summarized as:

- Setbacks to the development of small and large herds due to the drought's effects on the productive capacities of animals.
- Reduction of weight and hence marketable volume of meat because of under-nourishment.
- Reduced marketability of animals and their products due to the mal-nourished appearance of animals and the feeling (real or imagined) that they were afflicted by disease brought on by the drought.

- Reduction in the supply of local meats, creating more dependency on imported meat products.
- Significant income loss to livestock farmers.

Absence of Drought Insurance

The economic impacts on agriculture from drought were compounded by the absence of drought insurance for crops and livestock. Insurance is lacking for three key reasons:

- Insurers do not perceive this as a feasibly investment of resources, given the relatively small sizes of farms and livestock herds.
- Even if crop and livestock insurance for drought was available, small farmers most likely would not have seen the need. With the small size and subsistence nature of many operations, paying insurance premiums would have been considered an additional financial burden.
- Crop insurance is not integral to the small farm culture in Antigua/ Barbuda.

In the absence of being able to file claims for crop damages and/or loss, farmers had to absorb their own financial losses. The experience was difficult and stressful for many farmers and in some cases it dampened the enthusiasm for future investment in agriculture.

TOURISM

Landscape, Product and Climate

Antigua/Barbuda's tourism is promoted mainly on the strength of the quality of its beaches, climate and sea. Although not featured prominently in promotional literature, the scenic qualities and the overall attractiveness of the landscape are also critical to an enjoyable visitor experience.

Drought adversely affects the quality of the biological landscape (that is the living landscape of plants and animals) and the physical (non-living) landscape; the latter due mainly to the heat and dust associated with extended periods of low precipitation.

The general quality of the visitor experience was affected by relatively hotter and drier conditions, a degraded landscape and a less friendly disposition of residents under the stress from the drought. Uncomfortable and unattractive environmental conditions would have impacted negatively on decisions to take tours and hikes.

While these impacts may not have caused tourists to shorten their vacation, they could possibly have impacted negatively on decisions to make return visits.

The potential future income lost from return visits or from potential visits from friends who might have been influenced to visit if their experiences were more positive, cannot be determined. Failure to quantify economic loss in this case, however, should not reduce the significance of drought effects on the tourist product.

Water Shortages

The hotel sector, like all other sectors was affected by water shortages. Tourists are known to be highly intolerant of water deficiencies. Apart from the inconvenience, the implications for hygiene and health figure prominently in how they value the vacation experience. Although priority was given to hotels in water rationing, all properties were reportedly affected. Most properties lacked back-up water storage facilities of minimum standard size and even where such facilities were adequately sized in accordance with building standards capacities were irrelevant in the extended period of very low rainfall.

The economic effects from the drought on hotels resulted from:

- Shortening of vacation because of water problems and related inconveniences.
- Cancellations by potential guests because of the water problems revealed to them by travel agents or the media.
- Cost for purchasing water.
- Loss of income from all of the above.

Impacts on Property Landscapes

Properties reported damage and in many cases loss of ornamental plants. Such losses would have been more pronounced in exposed areas and in rapidly drained soils. The latter would include the sandy soils of some coastal properties.

Loss of Revenue to Hotels

Hotels incurred revenue loss. It would appear, however, that records on actual amounts are not available. Because of changes in hotel management since 1984, reliable estimates on revenue loss could not be obtained. The existence of the Crabbs desalination plant in the more recent drought in 1995 effectively reduced impacts related to domestic water supply.

Effects on Workers

The income loss to hotels also translated into hardships for hotel workers. Some were laid off during the period.

Restaurants

Drought impacts also extended to the restaurants. Restaurants serving local cuisine were affected by shortages in agricultural produce and meats. They were also affected by water shortages.

Settlements

The effects of the 1983-84 drought on households was as severe and, in some cases, more so than hotels. Many houses went without piped water for weeks at a time.

Drought conditions in 1983-84 affected:

- Household income
- Sanitation and Health
- School and work

(Water was actually barged from Dominica during the more severe periods of the drought and used to augment supplies to homes and hotels).

Household Income

Households where the main income earners were farmers (crop or livestock) experienced severe cuts in income and suffered from shortages in food. Generally, the reduction in crop yield, along with deteriorated livestock health and livestock mortality adversely affected food supply nationally.

Sanitation and Health

Extreme water shortages, such as those that occurred in periods of the 1983-84 drought, impact hygiene and sanitation, which in turn affect health. There are however no available reports on the impacts on the health of the population from the drought.

FUTURE VULNERABILITY

Agriculture and Crops

At the time of the 1983-84 drought, Antigua relied on surface water reservoirs to provide most of its potable water. Water supply was readily affected by meteorological drought and the country was highly vulnerable during periods of low rainfall. Desalination now accounts for 62 per cent of potable water and generally households and hotels have become significantly less vulnerability to deficiencies in rainfall. However, vulnerability remains high for crop farming because many farmers perceive the cost of water as one of the factors working against profitability.

A heavy reliance is still placed on rainfall to provide the moisture needed by crops, even when water from mains is available for irrigation.

Variable climate and precipitation remain over-riding critical natural factors contributing to the vulnerability of crop farming to drought. Practices (human factors) contributing to vulnerability include:

- Poor crop location in relation to soil and other factors that restrict plant growth and crop yield. Failure to match crops to land capability remains a problem. OAS produced land capability maps for Antigua and Barbuda in 1992. These should be used as the basis for building the awareness of farmers for better crop selection, along with improved soil and water conservation practices.

- Cost of irrigation.
- Water conservation practices –mulching is being used at Claremont Farms to conserve moisture in the growing of pineapples. A micro-irrigation (drip) system is used to conserve water. This is one of a growing number of examples where water conservation is being practiced. Unfortunately, most farmers do not seem inclined to use such conservation methods. The cost of mulch (organic or plastic) and the cost of water saving irrigation systems may act as deterrents to better conservation practices.
- Soil conservation practices – mulching not only conserves moisture but also protects soil from the forces of erosion. Soil particles do not bond adequately in the absence of soil moisture. As a result, erosion accelerates during periods of drought. Sloping terrain and areas exposed to wind are mostly affected. Nutrient loss accompanies erosion, so that crop yield is affected. Vulnerability can be reduced if constraints to wider use of mulch by farmers are removed.
- Absence of drought insurance –Farmers will continue to shoulder financial losses resulting from drought, because it is unlikely that local insurance companies will provide drought insurance for crops in the near future. Consideration should however be given to the design of a drought relief program by government, which could seek to access donor funding to assist farmers.

Livestock

Overgrazing. Overgrazing is perhaps the primary contributing factor to Antigua and Barbuda's vulnerability to drought. Reducing vulnerability in this respect requires drastic reduction in the goat populations and rigid control over where livestock are allowed to graze. To achieve these aims without causing unbearable economic dislocation and hardship to livestock farmers will require strategic planning, broad-based consultations with stakeholder groups and public awareness to build public support for actions to be taken.

Such actions should evolve from a national program and campaign to reduce the impacts of goats on the landscape, environment and crops. The program should have two primary components:

- Population control – seeking to achieve population grazing density limits that can be sustained.
- Restoration of forest or landscape areas that are displaying desert-like characteristics due to overgrazing.

Adequate resources would need to be applied for research, institutional building, public awareness and policy formulation in the implementation of the program.

Policy Issues

Reducing vulnerability to drought for both crop and livestock farming will require a number of critical policy initiatives, including:

- Incentives to achieve better soil and water conservation practices.
- Reduction in water rates for irrigation, through subsidy or other means.
- Development of adequate institutional infrastructure and coordinating mechanisms for drought planning and management, drought preparedness and drought warning. (This is relevant to other sectors, also, including tourism and settlements).
- Support for drought related research, monitoring, planning and mitigation.
- Regulation of livestock access to public owned lands.
- Stiffer penalties for livestock owners for damage to crops and gardens.

Tourism

After the 1983-84 drought a number of hotels installed reverse osmosis (RO) plants and have become less vulnerable to droughts as a result. Most hotels, guesthouses, villas and other forms of visitor accommodation properties still depend on APUA for most of their potable water supply.

Although the existence of the desalination plants makes water supply more dependable, properties without the capacity to produce most of their potable water are still affected by rationing. The extent of their vulnerability could not be determined in this assessment because a detailed survey of on-site water infrastructure, including production plants, roof catchment and storage capacity was not possible.

Based on discussions with a number of hotel managers, water conservation practices by hotels are limited to the use of water saving toilets and showerheads, which have become standard for many manufacturers of bathroom fixtures.

A lot of water is believed to be still wasted on properties due to leaks, poor irrigation practices, absence of a reliable level of wastewater treatment to allow reuse for landscaping and failure to meter water for different major property uses so as to detect waste. An inventory of tourism properties, including marinas and golf courses, should be considered to determine their preparedness to deal with future droughts. The inventory could be done with the involvement of the Antigua Hotels and Tourism Association (AHTA) and should cover among other things:

- Sources of water supply
- On-site water production capacity where appropriate
- Onsite water storage
- Irrigation systems

- Water conservation practices
- Methods for detecting water loss from leaks and faulty equipment

Settlements

Antigua

Like hotels, water supply to settlements has improved with the use of desalinated water by APUA. Vulnerability has been reduced with respect to water supply but most settlements are still affected by water rationing during droughts. This is expected to continue in the future unless APUA decides to increase water production from desalination on its own or with the use of private contractors. The relatively higher cost of producing desalinated water, compared to ground and surface water, is a major constraint to providing water that would not be affected by variable rainfall. High cost notwithstanding, a commitment to increasing desalinated water production may be required to reduce the vulnerability of settlements as water demands increase in the future.

In Antigua, water loss from evaporation and seepage at surface reservoirs is an issue without a cost-effective solution. In Nevis, where potable water is stored in tanks, such loss from evaporation and seepage does not occur. Despite the need to ration water in droughts, water is frequently lost due to leaks from supply mains that APUA is often slow to repair. The occurrence of this in the future will increase the country's vulnerability to drought.

In Barbuda, a major ground water aquifer has been affected by sand mining activity. Land and resource use will need to be managed in the future to ensure that this key water resource is not further compromised. Finally, although most building supply stores carry water saving toilets, showerheads and faucets, there has been no national water conservation campaign that targets households. New homes should be provided with incentives to install water saving fixtures and old households provided similar incentives to retrofit fixtures in order to conserve water. This would lessen the impact of drought on households.

Barbuda

Although a large percentage of households in Barbuda are without mains connection, vulnerability to drought is reduced by extensive nature of ground water resources on the island. A relatively high water table (3-4 ft. in some parts of Codrington) allows households to use wells to supplement rain or mains water. However, water from wells is often brackish and they risk contamination by virtue of their location relative to septic tanks and soakways.

Water quality is therefore a critical issue particularly in periods of drought when cistern water is not available for drinking and cooking. The use of a reverse osmosis (RO) plant to ensure a more reliable supply of potable water for Barbuda was proposed by APUA in 1996 but rejected by residents because wastewater from the plant would have been dumped in the Codrington lagoon (J. Mussington, pers. comm..). Desalination remains an option to be reconsidered in seeking to reduce future vulnerability to drought. Desalting brackish water should be less costly than using sea water, while the further exploration of ground water reserves could provide a cost-effective source of fresh water for future use.

Institutional Arrangements

Institutional capacity to manage drought is a critical factor in reducing vulnerability to drought events. Key agencies dealing with drought include APUA, the Department of Agriculture, the Livestock Division, Ministry of Planning, DCA and the National Office of Disaster Preparedness (NODS). The approach to planning, research and data management with respect to drought is ad hoc. The correct orientation and a commitment to drought impact mitigation, including adequate resource allocation and training of selected staff, are required.

Procedures and Information

It is recognized that drought is a recurring event in Antigua/Barbuda. However, apart from actions being taken to reduce problems related to potable water, there are no clear procedures in place to mitigate drought impacts and drought information management is disorganized. To improve the capacity for drought management, drought effects must be documented and a database to assist with decision making is needed. Also, procedures for drought awareness and building of public support for drought mitigation are critical measures to be considered.

ANALYSIS OF AREAS AT RISK

UNDERSTANDING AND APPLYING THE CONCEPT OF RISK

Drought risk for the purposes of this analysis is considered as having two (2) contributing factors, namely:

- Recurring environmental or meteorological events or natural physical characteristics beyond the control of man.
- Human practices (policy or land use) that make the population vulnerable to natural events and characteristics.

Natural risk factors

- Variable climate
- Exposure to wind and marine influences
- Exposure to excessive sunlight based on orientation to sun
- Problem soils, resulting from geological history, that may be soils

that are shallow, stony, erosion prone, have poor infiltration, heavily leached or lacking in nutrients

Land use practices leading to vulnerability

- Removal of forest cover, for agriculture, housing or other needs
- Overgrazing (high density livestock populations)
- Cultivation on steep slopes
- Disregard for water and soil conservation practices
- High per capita water consumption;

Policy

- Subsidized water costs
- Government land sales policy and subsidized infrastructure
- Lack of drought insurance
- Lack of drought relief programs.

AREAS AT RISK TO DROUGHT

Drought Risk Criteria for Mapping

Mapping of areas on the basis of their risk to drought – namely, low, moderate, high, very high – is based on a set of criteria. Each criterion is given a value of one (1) and the total value of an area determines its rank:

Low drought risk	<4
Moderate drought risk	5-6
High drought risk	7-8
Very high drought risk	>9

Drought risk criteria used for mapping are shown in Table. A description of the criteria follows:

Meteorological/Environmental

< 40 inches Rainfall Areas that experience < 40 inches of rain annually are meteorological disadvantaged not withstanding the wet periods of the year when soil moisture may be in excess of plant needs. For most of the year plants can be expected to experience moisture stress.

Exposure to wind and marine influences Areas with excessive exposure to wind and marine influences are determined by location, topography and slope type, using the PGDM base map with contours. These areas were noted but not mapped in assigning numerical values for drought risk ranking.

Areas on the east coast of Antigua and Barbuda are relatively more exposed to prevailing winds and hence more prone to wind erosion. Where they occur close to the sea, salt in atmospheric moisture retard the growth of most plants, except those that have become naturalized or adjusted to such conditions. However, vegetation recovery on cleared or overgrazed slopes then becomes very slow, irrespective of plant adaptability. When this happens vulnerability to drought increases. East facing convex slopes close to the sea are particularly prone to wind erosion and desertification if overgrazed.

Table. Drought Risk Criteria

Environmental Meteorological	
• Rainfall < 35 inches	1
• Exposure to wind and marine influences	1
• Shallow soils	1
• Slopes >11°	1
• Cactus scrub vegetation	1
Hydrological/Infrastructure	
• Absence of wells	1
• Absence of agricultural reservoirs	1
Human/Landuse	
• Crazing	1
• Crop location	1
• Population density > 5,000 per sq. mile	1 1

Shallow Soils. Shallow soils have limited moisture retention capacity and are often lacking in nutrients. Root development is often retarded because of poor soil structure. When such areas lose forest cover by clearing and are subjected to heavy grazing vegetation recovery is very slow and in some cases never occurs. This results in a trend to desertification. For Antigua, areas of shallow soils coincide with moderate to steep slopes.

Slopes > 11?. These are in Antigua and not Barbuda. Where such slopes have retained forest cover particularly moist forest or dry evergreen woodland, they are less prone to erosion and better able to retain moisture than slopes that have lost natural vegetation cover. Habitat degradation from drought is less severe on slopes in the former category.

Cactus Scrub Vegetation. For Antigua and Barbuda, the cactus scrub designation includes areas that were formerly cultivated and are now abandoned to extensive livestock grazing. The lack of dense vegetation cover in many parts of this vegetation zone encourages grazing and browsing by sheep and goats, major factors contributing to the vulnerability of the landscape to drought. Such areas visually exemplify the brown, harsh and oftentimes depressing conditions of droughts.

The Cactus Scrub designation, as used on the vegetation maps for Antigua and Barbuda, includes remnants of naturalized vegetation, which evolved on account of meteorological and other environmental influences, along with vegetation in various stages of succession after being cleared or cultivated. The former may include combinations of Acacia spp, lucaena and other legumes, agaves, cactus and other xerophitic species, all well adjusted to drought conditions. This type of association is considerably less vulnerable to drought except where coverage is sparse enough to allow overgrazing or browsing of the under storey. Where this happens, soil erosion or compaction can adversely affect soil fertility or otherwise retard plant growth.

Vulnerability is therefore more associated with Cactus scrub areas, where vegetative associations are less adopted to drought conditions or where land uses create imbalances in the landscape that may lead to severe erosion, desertification or flooding.

A weakness in using this criterion is the lack of adequate vegetation information that would allow a more detailed classification of the various types of vegetation associations found within the Cactus Scrub designation. Work is therefore needed to update, detail and more comprehensively map vegetation for both islands.

Notwithstanding the weakness in vegetation data and the drought resistant nature of many plant species found in the vegetation zone, Cactus Scrub areas remain vulnerable to drought for several reasons:

- Goats appear to like the general dryness of this type of landscape and appear to favor a number of herbaceous plants and other shrubs found on it.
- Where drought resistant species (example, Acacia spp., Logwood, Black Pearl) are established with sparse vegetative wood cover, over- grazing of under storey vegetation leads to erosion and slower plant re-growth.
- Where grass offers adequate cover for soil and moisture during normal conditions rapid loss of biomass due to water deficiencies in the soil and from grazing leaves the soil exposed.
- Extensive grazing in Cactus Scrub areas adversely affect plant diversity and the ecology of some areas; species liked by goats and sheep have difficulties surviving, while an invasive plant such as Neem (toxic to goats) becomes dominant in the landscape.
- Ornamental plants introduced in gardens falling in this general vegetation category are often not suited to local conditions – for example, water loving plants cultivated in areas with low precipitation. Such plants die quickly where water is not available for extended periods.

Exposure to Sun – This was also considered as a criterion but was not used for mapping drought risk because it requires very site-specific analysis of an area's topography, contours and vegetation to determine how these affect the amount of shade it enjoys during sunlight.

Concave slopes facing north or south enjoy relatively longer periods of shade during the day than east and west facing slopes. This may create slight but critical temperature differences amenable to moisture retention, vegetation and crop growth. This criterion can therefore be applied where detailed analysis of slope orientation to sunlight and shade is possible.

Hydrological/Infrastructure

Absence of wells. These are areas with no wells or known significant ground water deposits. In the absence of wells untapped ground water resources could be exploited with hastily drilled wells to offer relief in extended periods of drought. Information to map aquifers was not obtained. Water production information by wells in various watersheds was obtained and used in drought risk ranking for Antigua.

Absence of agricultural reservoirs – areas with no agricultural dams or not enough ponds available to livestock farmers. Agricultural reservoirs or ponds may run dry in an extended drought but nevertheless provide a critical source of water for farmers as long as it is available. Municipal and agricultural reservoirs and ponds are mapped on the watershed data map for Antigua.

Human/LandUse

Grazing. Grazing densities of > 6 goats or sheep per hectare or 1.5 cattle per hectare for pasture areas without irrigation are not considered sustainable. It is common knowledge that there are areas in the country that are excessively grazed. Although figures on the livestock populations for Antigua and Barbuda are available, population densities by areas are not known.

Where livestock densities are in excess of sustainable limits in improved (managed) pastures, impacts are mostly felt within the boundaries of the fenced pasture. Vulnerability from grazing is therefore due mainly to the threats to crops, gardens and habitats posed by un-tethered livestock found in Mixed Scrub and Grazing and Rough Grazing categories delineated in the land use data maps for Antigua and Barbuda.

Livestock population and grazing density data by area is required to strengthen vulnerability and impact analysis.

High Population Densities >5,000 per square mile. St. John's is the only settlement with this level of population density. Vulnerability linked to density derives mainly from limited space and poor soil conditions that limit storage capacity above or below ground.

Households built on small lots in parts of Point, Gray's Green and Ovals are particularly vulnerable. Such households are dependent mainly on piped water (attached to the house or available at stand pipes) and are unable to adequately store water when it has to be rationed.

Crops Location. This criterion targets areas located > 1 mile from mains supply. This applies to water loving crops in the volcanic region of Antigua as well as crops with lower water requirements grown in dryer areas. The latter is more drought resistant but vulnerability increases where there is no access to water supply mains.

Mango, which constitutes the major tree crop in the country, is less prone to damage from soil water loss than most perennial crops but is vulnerable nevertheless.Since the information to map water mains could not be acquired, the relative location of crops to main and secondary roads was used to assume their relative distance from water mains. Water distribution information, including variations in water pressure along the distribution network is required to reinforce the use of this criterion and to adjust drought maps as appropriate.

Crops are also vulnerable where their location relative to grazing activity may render them susceptible to damage from livestock, particularly during drought, when feed material becomes scarce. Denuded Slopes. Country wide mapping of denuded slopes was not possible during this assessment due to time constraints and therefore the criterion was not used in drought risk ranking. However, some of these areas are included in rough grazing land use designation on the land use data map for Antigua.

Such areas should be specifically mapped in the future since they provide ready evidence of how poor land use practices can create vulnerability to drought. Denuded slopes are areas where vegetation is almost absent and where most of the topsoil has been lost. They are indicative of a process, which started with the clearing for agriculture or firewood but have not recovered due to a combination of factors, namely, accelerated erosion, persistent and intense grazing and exposure to wind and marine influences.

Such areas exist on the southeast of Antigua between the Indian Creek area and Christian Hill. Some of these areas (slopes and associated valleys) are undergoing a process of desertification, which can only be reversed if the contributing landuse practices are curtailed. Such practices render the areas vulnerable to meteorological agents that hasten land degradation.

AREA AND ISSUES FOR PRIORITY ATTENTION

Areas

The east, north and southeast of Antigua contain areas of high and very high risk to drought. Environmental and land use factors comprise to make the region of the country the most vulnerable to drought. Within this region, the area requiring the most urgent attention falls within watershed 27-46, one of two watersheds ranked as being at very high risk to drought.

The area is at the extreme southeast of the island (between English Harbour and St. James Club). Environmental and human (landuse) factors combining to make this area highly vulnerable to drought include:

- Low rainfall
- Shallow soils
- Slopes exposed to wind and marine influences
- Overgrazing – the area has one of the highest goat population

Denuded slopes and narrow valleys in areas forming part of the Nelson's Dockyard National Park provide visual images of a severely degraded landscape and imminent desertification.

Similar trends are being experienced in other watersheds but perhaps not with the same degree of devastation. It is suggested that watershed 27-46 be considered for undertaking a pilot project seeking to provide strategies for reducing vulnerability and risk to drought.

The project should include:

- Forest monitoring program, particularly for cactus scrub and dry forest areas subject to threats from overgrazing. One objective would be to determine carrying capacity for livestock grazing and livestock population densities that should be allowed.
- Ecosystem studies aimed at better understanding the critical relationships between dietary requirements of goats and sheep, soil and moisture needs of preferred plant species and other ecosystem characteristics critical to understanding how plants and animals respond to drought.
- A landscape restoration component aimed at reversing trends towards desertification.
- The development of methodologies and approaches to involve livestock farmers in mitigating livestock impacts.
- The development of policy initiatives to reduce vulnerability and risk to drought.

Successful results of the pilot project would be replicated in other parts of the country. In Barbuda, the Palmetto Point, parts of the highlands, Codrington lagoon and the northern section of the island with Salinas, lagoons and mangroves are considered of low to moderate risk to drought. Most of the remaining lowlands are considered of high risk. No part of the island was categorized in the very high drought risk category.

Major issues

While the pilot project is being used to test approaches for mitigating the impacts of drought, actions should be considered and implemented to address other issues, including:

- Overgrazing. This is perhaps the singly most critical factor contributing to the country's vulnerability to drought. Trends in vegetation, habitat and landscape degradation, key factors contributing to the country's vulnerability to drought, will not be reversed until overgrazing is controlled.
- Meteorological data. A nationwide network of meteorological stations is needed collect data and monitor rainfall, evapo-transpiration, relative humidity, etc., to improve the country's database for informed analysis and decision-making.

- Vegetation classification and mapping. Existing vegetation data is inadequate for drought management. A vegetation classification and mapping project should be undertaken to improve the capacity required for managing various vegetation zones against drought impacts. Classification should include a description of soil and moisture requirements of key plant species within each vegetation class.
- Agricultural water storage infrastructure. The mapped data indicate an imbalance in the distribution of agricultural storage reservoirs and ponds. Vulnerability to droughts could be reduced for livestock and some crops by investment in additional reservoir and pond capacity in selected watersheds.
- Potable water. Water rationing by APUA in February and March, 2001, indicates that desalinated water (62 per cent of the total water production of 4.5 million cu.m/yr) has reduced but not eliminated vulnerability in respect of potable water supply. Since rainfall does not affect water production by desalination, rationing is necessary because of limitations or constrains in the supply of ground and surface water, which account for 33 per cent of total production. Additional water production capacity and/or storage are required if vulnerability is to be reduced in the future. This will be more so if significant growth occurs in the hotel and commercial sector, which now accounts for 17.5 per cent of current water consumption. The same will be true if major growth occurs in domestic water consumption, which currently accounts for 30 per cent of current total consumption.

DROUGHT MAPPING

DATA COLLECTION AND SOURCES

Data for drought hazard mapping were obtained from:

- Past projects and project reports, e.g. the OAS Natural Resources Assessment, Application and Projects for the Agricultural Sector of Antigua/Barbuda.
- Government agencies, such as the Coolidge Met Office, Ministry of Planning and the Development Control Authority (DCA). In the latter case, maps from the Draft National Physical Development Plan were used.
- Field observations.

Relevant sources are credited on individual maps.

Structure and Content of Maps

The drought hazard maps for Antigua and Barbuda were produced by analyzing ArcView GIS generated data maps with the following titles:

- Mean Rainfall Isohyets (Antigua)

- Vegetation (Antigua and Barbuda)
- Slope (Antigua and Barbuda)
- Watersheds (Antigua and Barbuda)
- Soils (Antigua and Barbuda)
- Land Use (Antigua and Barbuda)

The structure (content) of the each data set is given in Appendix 5.

Analysis of Data

Data was analyzed using ArcView. The Risk Criteria, identified in Section 4.0, were used as vulnerability themes and spatially manipulated by:

- Analyzing the spatial occurrence of vulnerability themes in the various watershed units. This was used to rank Antigua's watersheds in relation to drought risk based on numerical scores given in Appendix 6 (a). A similar system was used for Barbuda despite the fact that not all watershed boundaries are fully defined. Numerical scores for Barbuda's watersheds are given in Appendix 6 (b). Acreage data for land covered by relevant themes, e.g. grazing was not calculated but could be done in the future when vegetation and other data are updated;
- Overlaying (intersecting, as opposed to merging in ArcView GIS terminology) the themes to determine areas of critical risk.

By overlaying themes from data sets, critical areas within the watersheds that require remedial policy and land management actions can be identified.

Table. Drought Risk Ranking of Antigua Watersheds

Level of Risk	Watersheds
Low	1, 2, 3
Moderate	4-11, 12-20, 21-26,
High	47-53, 54-62, 63-66, 78-84, 85-86
Very High	27-46, 67-77

Table. Drought Risk Ranking of Barbuda Watersheds

Level of Risk	Watersheds
Low	I, X
Moderate	V, VI, VII, VIII
High	II, III, IV, IX
Very High	--

MAP USE AND LIMITATIONS

Map Use

The maps can be used for sector, strategic or development planning, disaster mitigation planning and drought management. The watershed, as a hydrological unit, with a common drainage basin is considered the most appropriate physical space for managing drought even though precipitation levels may vary within its boundaries. Hence, the ranking of watersheds in relation to drought risk provides the necessary background for further investigation and mitigation of drought impacts. For any given watershed, themes from data sets (maps) can be over-laid or merged to highlight actual or potential impacts from drought, as shown in the following examples: Overlay: Cactus Scrub vegetation + Rough Grazing and Mixed Scrub/Rough Grazing. Result: Tells the extent to which grazing impacts on this vegetation zone. Areas that fall outside of where the two themes intersect are often characterized by dense Acacia and other woody or thorny species, dense enough to discourage foraging by goats and sheep. Mitigation strategies to protect such areas, which may be the least degraded of the Cactus Scrub associations, will require updating field data for more accurate mapping and/ or detailed reclassification of the Cactus Scrub vegetation zone.

- Overlay. Grazing + > 11 degree slopes + exposed slopes (east-coast, Antigua) + < 40 inches rainfall. Result: Indicates areas of actual or potential landscape degradation, where vegetation re-growth is retarded; also indicates actual and potential areas of soil erosion from wind and rain;
- Merge. Agriculture (crops) + rough grazing and mixed scrub/rough grazing. Result: actual or potential threats to crops from livestock.

Limitations in Use of Maps

The maps can be used as references to drought mitigation planning and drought impact assessment. Data inadequacies should however determine the extent to which maps can be used to make decisions without additional research and field observations.

Attention should be given to weaknesses in various maps, namely:

- Mean Rainfall Isohyets (Antigua). There are not enough rainfall stations in Antigua and Barbuda to generate accurate lines of equal rainfall (isohyets). The map for Antigua was adopted from sources that did not indicate how many rainfall stations were used in producing the isohyets. A rainfall map was not produced for Barbuda because the data was not available. An adequate number of rainfall stations for both islands need to be installed and data collected on a regular basis.
- Vegetation (Antigua and Barbuda). No detailed and up-to-date maps produced from vegetation studies are available for either island. The vegetation maps are produced using different land use mapping sources and very limited field observation. Vegetation maps based on thorough field investigations and an agreed classification system is critical to drought mitigation.

- Slope (Antigua and Barbuda). The slope maps provide general rather than detailed slope categorization. Detailed slope analysis can be generated by GIS from the PGDM contour map. This should be done where site specific assessment of drought impact and mitigation is required.
- Watersheds (Antigua and Barbuda). Watershed boundaries were verified by study of a DOS contour map for Antigua and are believed to be accurate for Barbuda. The boundaries of critical aquifers were not mapped because the information was not obtained. This should be plotted as the information is obtained and where possible an investigation made of the effects of land uses in watersheds on aquifer recharge.
- Soils (Antigua and Barbuda). The soil classification used for both islands is general but adequate for this assessment. More detailed soil maps have been digitized for another study in the PGDM project and can be used for specific area analysis.
- Land Use (Antigua and Barbuda). Land use for Antigua was updated by the DCA (1999), after the OAS (1992), and adopted with slight modifications for the Antigua land use data map. The Barbuda map was updated from the OAS, 1992. Land use is however undergoing continuous change in Antigua and should be updated as significant changes are observed, particularly where practices could lead to increased vulnerability to drought.

IDENTIFICATION OF DROUGHTS

This section summarizes and describes indicators that suggest approaching or actual drought conditions. These indicators are discussed under the following categories:

- Meteorological/Environmental
- Hydrological
- Agricultural
- Socio-economic

ENVIRONMENTAL INDICATORS

- Reduction in biomass production of common grass species. Early warning is wilting, as grass roots become progressively damaged by lack of soil water. Grass cover becomes patchier as fallen leaf debris is blown away by wind in exposed areas. Homeowners find no need to use lawn mowers.
- Leaf fall and litter in Forests. The Forestry Division uses the increase in leaf fall and litter (detritus) on the forest floor as one indicator of drought. No measurement of volume of weight is done. Since leaf fall varies between species, measurement would need to be correlated with plant associations.

In Cactus Scrub forests, xerophytes (plants growing in dry areas) with small hard leaves can decrease water loss when the area begins to get drier and will drop their leaves in prolonged drought.

- Damage to "indicator" plant species. The Ficus, planted as hedge or trees, normally stays green; its small and shiny leaves limit moisture loss and its wide spreading root system allows it to seek moisture and nutrients beyond its drip line. When the Ficus loses a lot of its leaves it is an indication that the country is in a period of prolonged drought.

Some types of xerophytes, (succulents, such as cacti and agaves) utilize stored water in dry periods without noticeable damage. In fact, such plants can survive for weeks when uprooted, so that signs of damage (dried leaves or broken stems) may be an indication of negative water balance resulted from extended drought.

Hydrological

Reduction in Ground Water levels. The APUA Water Division uses ground water levels at various well fields as an indicator of hydrological drought. The approximate average water levels of selected well fields in Antigua are provided in Table. The table also gives the approximate levels for each field that is indicative of drought conditions.

There is a time lag between meteorological and agricultural drought conditions and hydrological drought as indicated by ground water levels. The latter occurs later but the time lag cannot be accurately predicted in the absence of rainfall data that could be correlated with rates of pumping and recharge, for the watersheds in which the well fields occur.

Recovery from meteorological and agriculture droughts occurs in advance of the replenishment of ground water to average levels. Again, the data does not exist to accurately predict the gap in time between these events.

Table. Drought Indicator Levels at Well Fields, Antigua

Well	Ave. Level (ft. below grade)	Drought Level
Bendals	40	60
Cades Bay	25-3	45-50
Claremont/Old Road	25-30	45-50
Collins/Bristol	15-25	25-45
Follies	50-60	70-80
The Valley	30-40	50-60

Water level figures given by APUA are approximations. This leaves room for more precision in measuring water levels as they recede as the basis for making decisions on water rationing. Reduction of Water Levels at Municipal Reservoirs. Receding water levels at municipal reservoirs is another drought indicator used by APUA. However, it was not determined if specific indicator levels for the various municipal reservoirs were established. Indicator levels at selected reservoirs could be determined over a period of time by widening the scope of monitoring to measure consistently meteorological and hydrological data, including, rainfall, temperature, wind speed, evaporation and seepage. Water Rationing. APUA begins to ration water before ground water levels recede to critical points. In fact, the decision to begin rationing water appears to be influenced by perceived meteorological and agricultural drought conditions. Reduction of Water Levels at Agricultural Reservoirs. This is another indicator observed but not measured by authorities.

AGRICULTURAL

Livestock

Livestock gain water from drinking water, foraging plants and water molecules produced in the breakdown of food. They lose water through urine, feces and water evaporated through skin and lungs. Some livestock can reabsorb water in their intestines to produce dryer feces and can reduce water used to excrete nitrogenous wastes of the urine. As animals well adapted to drought conditions goats are presumed to use similar physiological measures to conserve water. Despite this, they and other livestock will show signs of stress during extended drought:

- Weight loss. Symptoms are reduced muscle density, visible ribs. Effects are more associated with tethered livestock. However, untethered animals foraging larger areas for food use more energy and may suffer similar weight loss.
- Higher incidence of disease.
- Unusual high incidence of miscarriage among pregnant females.

Crops

Indicators may include:

- Negative water balance as evidenced from wilting and ultimately death of plants in extended droughts.
- Unusually high incidence of diseases, as plants are unable to obtain needed moisture and nutrients from the soil.

Socio-economic

Socio-economic indicators include changes in water use practices by households and businesses:

- Storage of water in existing or temporary storage facilities as a result of water rationing.
- Reduction of water used for landscaping of household and hotel gardens.

- Regular trucking of water to meet water deficit due to rationing.
- Higher incidence of respiratory ailments due to excessive dust in a very dry landscape. Respiratory ailments related to drought are not recorded for Antigua and Barbuda but are generally considered a drought related health impact.

6

Drought Management for Pastoral Livelihoods and Policy

INTRODUCTION AND BACKGROUND

Pastoral communities in arid and semi-arid regions of Africa live with the expectation of drought. They continue to suffer, and arguably increasingly suffer, catastrophic losses of livestock (capital and savings) during drought. The impact of drought is particularly acute for poorer members of communities with smaller livestock holdings and less developed social support networks.

Pressures on resources in arid and semi-arid areas have been growing in recent years as a result of human and livestock population growth and at the same time loss of land and water resources to other uses (agriculture, forestry, wildlife reserves). In general, land-use systems increasingly make use of more of the available natural resources in non-drought times, leaving fewer resources to cope with drought conditions. Growing links to the wider economy, and the development of local services (especially health and education) have led to various changes in the priorities of pastoral households and some pressure for sedentarisation to allow access to these services. Pastoral communities are therefore changing and there is increasing interest in the accumulation of cash as well as livestock. School fees need to be paid and pastoral communities increasingly realise that access to cash reserves can assist households to cope with drought-related calamities. However, few financial services are available to pastoral households. There is a need to take these changes into account in the design of drought management systems.

A basic model of a drought management system, adapted to a pastoral context. This includes a distinction between mitigation activities to minimise the impact of drought on production systems and livelihoods, and relief activities for the welfare of those rendered destitute. Often, and ideally, mitigation activities can be triggered by early warning and contingency planning during the onset of drought, while relief is generally more appropriate at a later stage.

Longer-term policies for resilience relate closely to, and are necessary for, specific mitigation measures. Thus plans to guarantee drought-time access to specific grazing reserves must be developed in the context of general policy on pastoral land tenure, and the efficacy of emergency marketing interventions may be severely limited by a lack of marketing infrastructure and price distortions in end markets. So too there are complex interrelations between mitigation measures, relief and rehabilitation. Relief should ideally be targeted on particularly vulnerable sections of the population that cannot be reached by mitigation measures.

Many drought mitigation activities have been tried over the years in northern Kenya beginning in the colonial era when destocking (purchase) of vulnerable stock took place on a large scale through the activities of the Livestock Marketing Division (LMD). More recently this approach has been replaced by smaller piecemeal responses (often by NGOs) attempting to mitigate the impact of drought by the provision of assistance in the form of marketing, water, veterinary services etc. Rarely have attempts to mitigate been well co-ordinated at local or national level and the major response over the past 2 decades has been one of 'last resort provision' of food relief to affected communities. Few efforts have been made to improve the self-sufficiency of pastoral communities and to assist the development of community-managed drought mitigation activities.

A number of factors now combine to make possible a more comprehensive approach to integration of drought management into pastoral development. Pastoral area support projects such as the Dutch-funded Drought Preparedness Intervention and Recovery Programme (DPIRP) and the World Bank-funded Arid Lands Resource Management Project (ALRMP) in Kenya have succeeded in institutionalising to a degree responses to drought in Northern Kenya. Support to pastoral livelihoods involves a range of activities including the development of service provision, early warning, support to drought mitigation activities, and the co-ordination of relief responses. Recent experience with drought mitigation and management in Kenya, in response to the major drought of 1999-2001 suggests that there may be an alternative to the provision of costly famine relief. Early warning, co-ordination of the activities of different agencies and the capacity to respond early at the first signs of drought are all key to the success of mitigation efforts.

DROUGHT MANAGEMENT AND RELIEF

There is increasing concern about the social, economic and environmental cost of drought (food or famine) relief. Massive efforts both logistical and financial are required to prevent the loss of human life. This approach, often dependent upon bilateral donors and international organisations (WFP and others) while effective in achieving their aim of reducing loss of life often lead to dependency while contributing little to the sustainability of pastoral livelihoods. Provision of relief is now expected by pastoral communities which may be leading to increased sedentarisation during periods of adequate rainfall in anticipation of handouts during drought.

Sedentarisation places great constraints on the mobility of livestock necessary for environmental sustainability of grazing. There is need therefore for a new approach. In Kenya this is recognised by both government and donors and is reflected in the efforts of both as they strive to improve drought preparedness particularly in the northern arid and semi-arid lands (ASALs). The activities of the Arid Lands Resource Management Project (ALRMP) has been critical in this regard. Working with the government of Kenya through the Office of the Prime Minister they have been instrumental in establishing in 10 northern ASAL Districts:

- Drought Steering Committees (DSGs) made up of both government and NGO representatives to plan and coordinate drought planning and responses
- Early warning systems in each District to facilitate rapid responses
- Preparation of drought contingency plans
- Improved drought preparedness (contingency boreholes cereal stores etc).

Kenya has therefore an efficient drought management structure, which has proved its value by saving lives and livelihoods cost-effectively in the 1999-2001 drought. The issue is now how to move drought management forward by clarifying policy and institutional requirements. This paper looks at policy conclusions about costs and benefits of drought management, especially the Early Warning (EW) system, the maintenance of standby drought management capability, and rapid reactions. Only a small number of potential reactions are analysed, out of a much larger pool. The paper does not look at costs and benefits of preparedness measures, nor rehabilitation, although these are likely to have highly positive benefit to cost ratios.

Drought management and mitigation will be important for the future sustainability of pastoral livelihoods and will require the intervention of government and donors. The paper also examines ways in which pastoral communities can become less dependent by the development of financial services tailored to their specific circumstances.

POLICY OBJECTIVES

The overarching policy goal for drought management should be to minimise the impact on livelihoods and the general economic, social and environmental disruption which may be triggered by severe drought. This can be broken down into a set of policy objectives. The main ones include the maintenance during drought of:

- The purchasing power of vulnerable populations, especially pastoralists
- The physical availability of cereals throughout affected districts

- A dispersed population distribution (to reduce the risk of contagious disease in refugee camps)
- Adequate levels of human nutrition and health care
- Adequate levels of animal health care
- School enrolment
- Livestock and human water availability at widely dispersed water sources
- Security (minimising conflict)

The main methods available to government to achieve these policy objectives include developing an institutional capacity to:

- Generate accurate community, district and national level early warning and monitoring data
- Intervene early in drought crises
- Manage drought responses and rehabilitation
- Promote subsidiarity in drought management - actions should be decided and managed at the lowest feasible level
- Make available flexible contingency funds at district level for rapid implementation of drought responses.

Public and Private Goods

Many of the policy objectives listed above properly concern public goods, which are the responsibility of the government to provide. In the strict economic sense of goods that are nonexcludable (it is difficult to exclude anyone from their benefits) and nonsubtractable (enjoyment by one user does not detract from enjoyment by others), public goods include: adequate drought early warning and management capacity, peace and security, some animal and human health protection (e.g. against highly contagious diseases such as rinderpest, measles). However, there are broader senses of the term: that without external assistance poor pastoralists could not provide themselves with such goods, and that the result of such market failures would either be intolerable suffering or later, and more costly, provision of relief. In these senses public goods would include curative aspects of human and animal health, the provision of water, and the maintenance of schooling through droughts. These definitions of public goods and the responsibilities of government should be the subject of public discussion.

Who Pays, who Benefits?

Provisional cost:benefit calculations of Kenyan drought mitigation projects, suggest that on overall figures (all benefits versus all costs) there are considerable advantages to many types of drought intervention. But in most cases costs are paid by donors (in some cases by government), and benefits accrue to herders (occasionally to government). This means that simple benefit:cost logic (if benefits exceed costs, do it) does not apply.

There are several ways to look at this. Firstly, as pointed out in the previous paragraph, many of the activities analysed here concern public goods: these should therefore be provided by government irrespective of the benefit:cost calculus. These figures merely show that the interventions concerned are well-designed and sensible, and provide an additional incentive to government to provide the public goods concerned.

Second, it is likely that donors will continue to support Kenya in droughts. In that case, the cost:benefit analyses provide one way to rank potential interventions on the basis of the relative impact of a given unit of aid expenditure. It makes sense for donors to support any project with a positive benefit:cost ratio, and particularly to support those with high ratios.

Third, and perhaps most interestingly, this analysis opens up the possibility, in the case of private goods, of the development of appropriate financial instruments to allow herders themselves to invest in some of the activities discussed here. If benefits substantially exceed costs, the issue becomes one of scheduling payments and receipts over time. Some imaginative micro-finance design might provide interesting ways forward.

METHODS FOR DROUGHT MANAGEMENT

Institutional Structure for Drought Management

Key elements of a successfully institutionalised drought management strategy, as already developed in Kenya, include:

- A unified system in which preparation, early warning (EW), response and rehabilitation are closely linked to each other and are under the same control
- An emphasis on district level authority and resources, and good district coordination
- Good links between districts and Nairobi
- Access for the head of drought management to the Permanent Secretary in the Office of the President (OP)
- Flexibility and speed of response, created in part by district contingency funds earmarked solely for drought

Early Warning and Preparedness

The first essential component of a drought management system is an early warning system linked closely to a standby capacity to respond to and manage drought. This already exists in Kenya through the Drought Management Office in the OP, the District Drought Management Offices in ten arid districts, and the associated institutional structures such as the District Steering Groups (DSG).

Early warning and the standby capacity to manage drought are important public goods, and it is part of the responsibility of government to provide it to facilitate government's own operations, those of donors and NGOs, and the efforts of local communities. Additionally, effective early warning and standby management capacity can be justified in narrow cost-benefit terms. The recurrent costs of maintaining an early warning and contingency planning system would be justified simply by allowing a 5.5 per cent efficiency gains in the delivery of food aid, if serious drought necessitating large-scale food relief occurred every five years. Such a gain would in fact be achievable through the better targeting that a planning system would allow, before even considering the greater gains possible through timely mitigation. Even if the analysis is limited to only Kenyan government expenditure on food aid, the EWs and management system only needs to generate a 16.5 percent saving in food aid to pay for itself.

Rapid Reaction

An efficient early warning and management system also makes possible a range of interventions to mitigate drought once it occurs. Such interventions provide both public goods (such as preventive animal health) and private goods (such as animal dewormers and supplementary feeds). They may also contribute to the fight against poverty. Many are also cost-effective in narrow economic terms in the sense that the immediate economic benefits outweigh the costs (without considering to whom benefits and costs accrue).

Many of the most important benefits from better drought management are for practical purposes unquantifiable. Although we cannot realistically quantify these benefits, there is no doubt they are substantial, and they must not be forgotten in any calculation. On the data available, it is also difficult to calculate overall costs and benefits of many theoretically quantifiable interventions in past or future droughts. But it is important to be aware of the scale of losses caused to the livestock sector by drought, against which some sense of the benefits of mitigation can be evaluated.

Economic losses during drought

In a drought pastoral households sell animals in order to buy staple cereals. Because everyone is selling, and there are few buyers, prices fall substantially. If the animals have a buyer, this does not represent a loss to the overall Kenyan economy, but their low value represents loss of a key capital asset to the seller household, and a common reason for household descent into poverty. Helping them out of poverty will in the future carry a significant cost.

No figures are available for the level of distress sales of livestock during the 1999-2001 drought or of the prices received by pastoral households. The general experience in droughts in other parts of Africa is that more animals are sold than die, at prices which are often half or less of post-drought prices. As a very rough indication of orders of magnitude, we may guess conservatively that as many animals were sold as died (30 percent of cattle and small stock populations and 18 percent of camel populations), and that drought prices were half post drought prices. On these assumptions, economic losses to households from distress sales would be the same as estimated mortality losses, or $US 77.3 million. As before these are for both northern and southern rangelands and somewhat overstate the losses for the ten arid districts.

Reducing Livestock Mortality

One estimate of 1999-2001 livestock drought losses for Kenya as a whole is 2.3 million sheep and goats, over 900,000 cattle, and 14,000 camels, valued at approximately $US 77.3 million. These are losses for northern and southern rangelands combined, so somewhat overstate losses in northern arid districts alone. Given the level of early warning and interventions, this level of mortality should be considered appropriate to a partially-managed drought, with losses larger than for a well-managed drought, but smaller than for an unmanaged drought. A comparable estimate for loss of animals in one district – Wajir - alone during the 2000-1 drought is $US 7.6 million.

Animal deaths are a total write-off of the value of the animal to the household and to the national economy, and the loss of a valuable production asset at both levels. In a pastoral economy, animal deaths threaten livelihoods. Preventing deaths not only protects a key asset, but also reduces post-drought expenditure on restocking.

Livestock Interventions

Rapid reactions in a well-managed drought can considerably reduce livestock mortality, through early and appropriate measures. These are likely to have the effect of:

- contributing to recovery as pastoral household purchasing power is maintained and can be used for purchasing drugs and feed to keep valuable breeding stock alive to aid herd re-growth post-drought.
- reducing human suffering and destitution
- reducing the need for costly food or famine relief,

Interventions for which tested models are available include:

- destocking (purchase of stock threatened by drought)
- appropriate animal health care
- improved grazing (including cross-border arrangements)
- secure water provision
- provision of supplementary feeds (in association with destocking)

It was estimated that during the six month period June 2000 to January 2001, donor livestock interventions funded by CDTF, DFID, USAID and others amounted to $US 4 million (Ksh 300 million). The largest number of projects during the period 2000-2001 in northern Kenya involved de-stocking followed by the emergency veterinary assistance. The water sector consumed the largest amount of resources.

	million $US
– destocking/offtake	0.98
– transport subsidy	0.12
- animal health	1.16
– livestock feed	0.14
– restocking	0.11
– cross-border grazing	0.07
– other (especially water)	1.5
Total	4.08 million

Aklilu and Wekesa estimated that the economic benefit (in terms of value of animals saved and livelihoods thus protected) from these operations was $US 10 million (Ksh 750 million). Some further cost:benefit analysis of particular projects, and discussion of different kinds of intervention, are given by Morton et al. 2002. The main livelihood threat to pastoral households in a drought is the erosion of pastoral terms of trade (the value of things pastoralists have to trade – determined by the number of saleable animals and their price – relative to the things they have to buy – mainly staple cereals). Where the market price of animals falls relative to the price of cereals, herding households will have to sell more animals to acquire the same amount of cereals. Animal prices commonly fall in a drought, since there are many sellers and few buyers, and herders are often obliged to sell more, and better quality, animals to meet household minimum food intake requirements.

Emergency purchase of livestock is therefore a key strategy to maintain livelihoods. It can be implemented either through direct purchase by implementing agencies, which has the benefit that meat can be distributed as a relief good or through schools, providing valuable protein; or through transport subsidies paid to livestock merchants to encourage them to purchase from more remote areas, from which it is difficult to trek drought-stricken cattle. The additional purchasing power that pastoralists gain from timely emergency purchase can be used to provide veterinary drugs, supplementary feed and water, thus raising chances of herd survival.

Other technically successful and economically productive interventions include veterinary service provision, and provision of emergency boreholes and borehole repair and maintenance. Water interventions may substantially reduce the time women spend collecting water for the household. This has far-reaching benefits for household cohesion and childcare.

Participation of Pastoral Communities in Decision-Making

Several reviews have been made concerning the participation of communities in the drought mitigation process. The general consensus is that although communities were generally consulted about which type of intervention they thought best suited there needs, in some cases women were not consulted. In other cases no community co-ordinating bodies were set up to deliver interventions, little training was offered and interventions were not always well co-ordinated with development priorities and objectives. It is generally understood that balanced gender/community participation in planning and implementing mitigation improves effectiveness and efficiency and responds to household priorities

Co-ordination and Management of Drought Mitigation

Institutional factors are the main impediment to timely responses to drought emergencies. These include the response time of donors and government to requests for funds for mitigation activities. In general EW is effective and Drought Steering Groups (DSGs) are therefore able to respond quickly, but the process is slowed down by the bureaucratic procedures of donors.

Having emergency funds in place at the District level for management and co-ordination by DSGs or other appropriate bodies could do much to speed up responses and improve mitigation. This has implications for decentralisation of power from central to local government.

A co-ordinated approach to planning and implementation of mitigation is key to its success. To a great extent this was achieved in Kenya in the 1999-2000 drought. National co-ordination was by the Office of the President and the KFSSG and the various special interest committees such as the Livestock Sub-Sector Working Group.

At the District level the role of the DSG was crucial and to a large extent local mitigation interventions were well co-ordinated. In some cases however, donors bypassed the DSG and began implementing activities in their own right. This led to some duplication and waste of resources, particularly in the field of water provision.

Emergency interventions place high administrative demands on implementing agencies and require a degree of management skills that may be beyond the smaller local NGOs. There is an opportunity to transfer the required skills to these organisations in non-drought periods.

The engagement of local NGOs as subcontractors in both livestock related mitigation activities and as agencies responsible for the delivery of food aid was considered to be an outstanding success. These organisations have local knowledge and connections with local communities. They tend therefore to be trusted by pastoral households. The level of decentralisation of relief and mitigation efforts provides evidence of the need for greater local (district) decision making when dealing with emergencies and emphasises the need for a local fund for rapid response.

PUBLIC POLICY AND DROUGHT MANAGEMENT IN AGROPASTORAL SYSTEMS

Agropastoral farming systems are well suited to drought-prone areas with low rainfall, where, despite harsh conditions, these systems have provided reasonable sustenance to pastoral societies for generations. Unfortunately, the levels of wealth accumulated in these societies is inadequate for providing full protection from severe droughts, and the economic and human losses in drought periods can be severe. The problem has worsened with population growth, as more and more people seek to earn a livelihood from the meager resources available in these areas. It may also have been aggravated by more frequent and prolonged droughts associated with global warming.

The high cost and the increasing vulnerability of agropastoral societies has led many governments and donors to intervene with various forms of drought assistance. Types and levels of intervention that were unheard of until recently are now common, and seem to be increasing. Many of these interventions are encouraging farming practices that increase both the extent of future drought losses and the dependence of local people on government assistance. They are also costly to governments and donors, and use resources that could otherwise be spent for development purposes. It is important to know if the net benefits from existing types of drought-relief programs justify their costs, particularly when their longer-term impacts on poverty and the environment are assessed. It is also important to know if drought-relief programs can be designed better to achieve their immediate objectives but without distorting economic incentives in inappropriate ways. This chapter addresses these issues.

NATURE OF DROUGHTS

Droughts are very common in most rangeland areas. These areas have low rainfall on average and high coefficients of variation, hence the probability of rainfall falling below "critical" levels for forage production necessary for sustaining herds is quite common. What is a critical rainfall outcome? Pratt, Le Gall, and de Haan (1997) suggest a drought can be said to occur when rainfall falls below half the long-term average, or when rainfall in two or more successive years falls 75 percent below average. Rainfall failure stunts pastures, desiccates water points, and kills livestock. It leads to the liquidation of a significant part of total herd or flock in the absence of other sources of feed. Moreover, since the main commercial output of pastoral systems is meat, prices for meat tend to be negatively correlated with drought (more animals are available for sale in drought years), which accentuates income shortfalls. As human populations grow, so do animal stocking rates; and pastures are put under increasing stress, which increases their vulnerability to drought. What used to be a manageable rainfall outcome may now be considered a serious drought that leads to significant economic and social costs.

How herders traditionally manage droughts

Agropastoral societies have developed their own strategies for reducing their exposure to losses from droughts and for coping when droughts occur. These strategies include

- Diversifying into crops and livestock, particularly around settlement areas, and diversifying into different animal species (for example, goats, sheep, cattle, donkeys, and camels) and different breeds;
- Carrying extra animals that can be liquidated easily during a drought, either for food or cash;
- Aadopting mobile or transhumant grazing practices that reduce the risk of having insufficient forage in any one location;
- Adopting opportunistic grazing practices whereby herd sizes and stocking rates are adjusted as the rainy season unfolds to best match available grazing resources;
- Maintaining reciprocal grazing arrangements with more distant communities for use in drought years;
- Maintaining feed reserves or purchasing supplementary feed, such as hay or concentrate;
- Investing in wells and cisterns; and
- Diversifying into non-agriculture, particularly seasonal migration and non-farm employment.

Traditional risk-management strategies have proved effective in managing drought risk and have enabled pastoral societies to survive harsh environments for many millennia. The interplay between drought and traditional management systems also helped keep total herd sizes in equilibrium with the inherent productivity of the pastures, avoiding any long-term degradation of grazing areas. Stocking rates would trend upward between droughts, as herders bred more animals, but then would be knocked down again when the next drought occurred. Fluctuations in herd size closely followed rainfall patterns, and peak stocking rates rarely reached unsustainable levels.

However, despite their advantages, traditional drought-management strategies can also have associated opportunity costs. Thinking in terms of two types of costs is useful: the opportunity costs arising from inefficient use of resources within existing agropastoral systems, and the opportunity costs arising from failure to exploit more productive agricultural development pathways that could change the farming system in fundamental ways.

The first type of opportunity cost may occur, for example, when herders liquidate animals during droughts and then end up with too few animals in the immediate post drought period—and hence miss out on important short-term production opportunities. On the other hand, given a sufficient respite between droughts, herders may build up excessive herd sizes to have a liquid asset as a hedge against the next drought. This can lead to overgrazing and the degradation of pasture, with reduced productivity. Herders also prefer to keep traditional breeds that are more drought tolerant but less productive. Herders also are often less willing to use or invest in modern inputs (for example, feeds and veterinary treatments) that could increase average profitability but lead to loss of capital investment if rainfall is unfavorable. Surprisingly, little quantitative information seems to be available about these costs, although they could be analyzed relatively easily with the aid of stochastic programming models of problems in household and community decision making.

The second type of opportunity costs are more speculative. If mobility and transhumant grazing practices remain the primary strategy for managing drought risk, communities must retain large areas of land as common properties and retain reciprocal grazing arrangements with other communities for use in drought years. This necessarily restrains the enclosure and privatization of land, which in turn can impede investment in land improvements and the development of more intensive and settled farming systems. Without a shift to such intensification strategies, it is not clear how communities can continue to absorb increases in their populations. One alternative is for local communities to manage grazing rights to common property more effectively, with collective investment in land improvements. However, the absence of many successful examples suggests that such collective investment is extremely difficult to organize, manage, and sustain—particularly in the context of rapid population growth and the increasing commercialization of agriculture, which make cropping increasingly attractive.

The lack of quantitative information about the opportunity costs of traditional risk-management strategies in agropastoral systems is a problem for the design of good drought-management policies, because their costs should be justified on the basis of the increases in productivity and incomes that they generate. Much more research is needed to determine just how elastic the productivity of pastoral systems are to changes in risk-management opportunities. Are there important opportunities for increasing production that could be exploited if drought risk could be more effectively managed?

Reasons for Public Drought-Management Interventions

If traditional drought-management practices are constraining growth, it is relevant to ask if the problem could be managed more efficiently. This is equivalent to asking if the government could correct market failures. How might such market failures arise? Two possibilities seem obvious: First, the covariate nature of drought risk makes more efficient risk spreading difficult within pastoral societies; everybody suffers when drought occurs, and local sources of credit dry up just at time when they are most needed. Second, livestock prices plummet during droughts, when everybody is trying to sell, and then rise rapidly afterwards when everybody is trying to buy to rebuild flocks at the same time. Credit and insurance markets for diffusing this covariate risk are weak in many drought-prone rural areas.

Property rights problems related to the ownership of crop and range land may prevent the spread of management practices and investments that lead to more efficient drought-management strategies. For example, incentives to plant shrubs or build up fodder reserves on range land that is communally owned but ineffectively regulated may be insufficient. Also, without adequate property rights, population growth can lead to excessive stocking rates, and to encroachment of the cultivated area into traditional rangeland areas. These changes in turn can induce degradation of range and soil and, by restricting the spatial mobility of flocks, increase herders' exposure to drought risk.

These kinds of market failures can provide a rationale for public intervention. However, governments might also be motivated to intervene with drought-management policies for other social and environmental reasons, too, including the following:

- Government has an obligation to alleviate human misery in drought years and to help protect the stock of breeding animals for the future. Drought-relief measures may be seen as cheaper than safety-net programs.
- Herders and farmers may default on loans in drought years, causing difficult problems for lending institutions.
- Overgrazing of pastures that are already drought stressed, and soil compaction in areas around water holes, may contribute to wind erosion and local climate changes that are negative externalities for a country.

Economic Aspects of Public Interventions for Drought Management

While humanitarian objectives are often the initial reason for government and donor drought-relief interventions, they can prove economically expensive in the long term if they are not designed to overcome some more fundamental problem, such as a market failure, that prevents more efficient and productive use of resources in agropastoral systems. Moreover, simply being able to fix an underlying problem is not sufficient to ensure that it is economically worthwhile. The problem must also be fixed at a cost that is less than the benefits, and in ways that give a reasonable rate of return on public funds.

Where drought relief is required as a result of market failures (for example, inappropriate property-rights systems, or a poorly developed financial market), fixing the underlying problem (for example, reforming property rights or strengthening rural financial markets) may be more efficient than incurring the repeated costs of drought relief. Similarly, public-investment opportunities for reducing drought losses (for example, water catchment areas and wells) may also be more cost effective over time than drought relief. Unfortunately, these kinds of opportunities are typically quite limited in many drought-prone areas, and public drought-relief programs may not be an option in the near term. The costs of public drought-management interventions are relatively easy to determine, but the benefits are much harder to assess. One-time interventions can provide significant humanitarian relief. However, once drought-management policies become institutionalized so that farmers and herders begin to take them for granted, these policies can lead to important changes in farming practices that affect productivity. Well-designed and -implemented drought-management policies can contribute to greater productivity and thereby justify their costs. However, poorly designed interventions may lead to small productivity gains or may even be counterproductive.

Any good risk-management aid should enable farmers and herders to take greater risks in their quest for higher average returns. If farmers are risk averse, they trade off some level of expected income for lower risk (for example, through diversification strategies). The amount of expected income forgone to reduce risk can be viewed as a risk premium paid, or a production cost.

If this cost can be reduced by the introduction of an improved risk-management aid, the farmer can change strategy (for example, specialize more in the most profitable activities) and obtain a higher average income for the same amount of risk. This change not only improves expected farm incomes, but can also lead to spillover benefits to consumers at an aggregate level through lower prices as the supply function shifts downward by the amount of the reduction in the risk premium per unit of output. This effect is very similar to the effect of a new cost-reducing technology and, providing the new risk management aid is not subsidized, a net gain in social welfare always occurs.

However, if the new risk-management aid is subsidized, the effect is similar to a subsidy on any other farm input (for example, fertilizer or credit). The reduction in unit costs is partly paid for by the subsidy, and the dead weight loss of the subsidy is always greater than the sum of the additional producer and consumer welfare that it generates. What does this mean in practice? It means that subsidized drought-management interventions reduce risk costs to farmers to below their true social value, leading to excessive risk taking and increased exposure to future drought losses. Not only is dependence on future drought assistance from the government built in, but the net social return to that assistance to the country can be small or even negative. The bottom line is that, wherever possible, public interventions should be limited to drought-management interventions that farmers pay for themselves. In the following section are examples of interventions that led to inappropriate and socially costly responses by herders. Another potential problem with poorly designed drought-management policies is that they can lead to moral hazard. This is a well-known problem in the insurance literature and refers to the incentive problems that arise when an insurer underwrites risks whose outcomes can be influenced by the insured's behavior. For example, if an insurance company contracts to compensate farmers for yield losses against pest and disease damage, farmers will have reduced incentive to be diligent in protecting or treating their crops once they realize that the insurance will compensate for losses anyway. Moral hazard leads to greater losses than necessary, increases the risk exposure of the insurer, and makes calculating those risks actuarially almost impossible.

Similar problems can arise if government indiscriminately compensates for drought losses that could be reduced or avoided by herders. Unless appropriately targeted, restocking programs could, for example, lead to less-diligent care of livestock during droughts, or even to fraudulent claims for supposedly dead animals that were in fact sold. Such behavior would lead to greater losses than necessary and make restocking programs more expensive than they need be. Similar problems can arise with feed-subsidy programs if the feed is distributed indiscriminately, leading to reduced incentive to exploit remaining grazing opportunities, particularly in more remote areas that require greater time and expense to reach.

Past Experience with Drought-Intervention Policies

Many governments have intervened to help manage drought losses, but usually on the basis of crisis relief once the drought has set in (for example, food for work, distribution of subsidized feeds for livestock, and assistance with restocking). Since the primary motive is typically humanitarian assistance, not much thought is given to the longer-term impacts of drought interventions on farming practices and productivity. This has been especially true of much of the crisis aid provided by donors and nongovernmental organizations. The result is often an inappropriate set of economic signals to farmers and herders, leading to unsustainable farming practices in many drought-prone areas that increase both future drought losses and farmers' dependence on government assistance, and to moral hazard that further add to the government's cost of providing drought compensation.

A good analogy is the experience with hurricane-disaster assistance in the United States. By routinely stepping in to compensate home owners for their losses after a hurricane, the government encourages home construction in vulnerable coastal areas where prudent investors would not otherwise build and encourages fraudulent practices within the home-repair and construction industry. These problems add enormously to the cost of government assistance over time.

Since this chapter is concerned with drought-management policies as an aid to more efficient risk management in agropastoral systems, it focuses on their longer-term impacts on resource management and productivity. Two types of interventions are reviewed: feed subsidies, which have been used widely in the West Asia and North Africa region; and restocking programs, which have become popular in many parts of Sub-Saharan Africa. Both are perceived as longer-term programs rather than as simply ad hoc relief (although they often began that way), and the body of evidence on their impacts is growing.

FEED SUBSIDIES IN THE WEST ASIA AND NORTH AFRICA REGION

Droughts have long been a significant factor in the West Asia and North Africa region, and particularly for the crop–livestock systems of the low-rainfall areas and the vast grazing areas of the steppe. The problems were severe back in biblical times, but as human and livestock numbers have increased over the centuries, the total magnitude of the economic costs caused by droughts has increased at least proportionally. Oram (1998) provides a recent review of some of the consequences of drought for livestock production in the region. In the 1945 Moroccan drought, for example, half of the national sheep flock died, and in the 1981–82 drought, 25 percent of the cattle and 39 percent of the sheep either died or were sold prematurely on a glutted market. In a major drought between 1958 and 1962, at least 70 percent of the then considerable camel herd in Jordan died, leading to a virtual demise of camels as an economic element in livestock production there. In Syria, some 3 million sheep (about 25 percent of the flocks) had to be slaughtered during the 1983–84 drought because of a shortage of feed.

To reduce these kinds of losses, governments throughout the region introduced extensive drought-management policies during recent decades. These interventions focused on providing supplementary feeds to safeguard livestock, with the predominant expenditure going for subsidies toward the costs and distribution (usually by parastatals) of concentrates and other feeds. Barley is the most commonly subsidized feed, and the extent of the subsidy extends to as much as 32 percent in Tunisia and 50 percent in Morocco. Feed imports are also relaxed in drought years, while imports of livestock and livestock products are constrained to maintain domestic prices.

These programs have been very successful in protecting livestock numbers and production during droughts. Although the 1995 Moroccan drought was devastating—with total cereal production falling to only 17 percent of that in the good 1994 season—the ruminant livestock sector was barely affected. In Tunisia, Boughanmi (1996) estimates that sheep numbers rose consistently during the droughts of the early 1990s, while the World Bank estimates that, in the absence of the drought relief measures, producer prices would have dropped by approximately 40 percent during the 1988–89 drought. According to the Bank, potential losses to producers during 1988 and 1989 could have been 119.7 million dinars ($133 million), and this was prevented by a program that cost 74 million dinars ($82 million). By preventing the large-scale loss of livestock, the interventions also avoided production losses in subsequent years, too, but these are not included in the World Bank's calculation.

Although they have helped limit production losses caused by drought, the drought-management programs have also had negative impacts:

- They have accelerated rangeland degradation in the long term by undermining the normal process of adjusting flock size to interannual climatic variations. Herd sizes have increased sharply in recent years, and grazing practices have changed so that many of the animals no longer leave the rangeland areas during the dry season but have their feed and water trucked in. This practice leads to overgrazing during the dry season, prevents the natural seeding of annual pasture species, and disturbs the soil and contributes to wind erosion, particularly in areas near water and feed supply-points. The degradation of the range and soil has been exasperated by high government procurement prices for barley, which has encouraged the mechanized encroachment of barley cultivation into rangeland areas where it cannot be sustained.
- They have added to the fiscal burden on governments. During the 1994 drought, for example, some 500,000 tons of heavily subsidized concentrate were fed to livestock under the program in Tunisia. Even larger amounts were distributed during the 1988–89 drought at a cost to the government of 74 million dinars ($82 million). In Morocco, an estimated 420,000 tons of feed were distributed during the 1992 drought—at a cost to the government of about $30 million. The same program cost the government about $28 million during the 1995 drought. In Jordan, cumulative feed-subsidy costs between 1991 and 1996 were 168.4 million dinars ($116 million), and $55 million of this was incurred in the 1996 drought alone.
- They have proved difficult to target, with the lion's share of the subsidized concentrates going to large herders and to commercial farms. Indeed, during the Moroccan drought of 1992, many farmers received only small amounts of subsidized feed—about 1 percent of total feed needs.
- They have a tendency to become permanent. Many West Asia and North Africa countries quickly moved to permanent feed-subsidy systems and now expend considerable public resources on the distribution of heavily subsidized concentrates every year.

The high budgetary cost and the negative environmental impacts have now led some countries to reconsider their drought-management programs. Jordan recently abolished its feed-subsidy program as part of its structural adjustment program, which has led to a very sizeable reduction in the national flock of sheep and goats (perhaps by as much as 40 percent in two years).

Syria maintains its feed-subsidy program but has now banned the cultivation of barley in many of the steppe areas.

Restocking Policies in sub-Saharan Africa

The primary objective of restocking programs is to help herders reestablish their herd sizes as quickly as possible after a drought. Traditionally, herders often sought to contract out the management of part of their herd in drought years to nomadic herders who could take the animals to other regions for the duration of the drought. Their return with the rains had much the same effect as a contemporary restocking program. The simplest interventionist schemes amount to little more than targeted handouts of live animals during the recovery period, usually to the most impoverished herders.

However, more complete approaches seek to provide a better balance between fodder availability and livestock numbers throughout the drought cycle, as well as to help stabilize livestock prices. This can be achieved by buying up animals at some reasonable price during the drought, when fodder is scarce, and then reselling animals back to herders at the same or slightly higher price once drought recovery begins. This approach is sometimes characterized as a livestock bank, in which animals are essentially deposited into a savings account for the duration of the drought. Difficulties can obviously arise in setting targets for the number of animals to buy or sell: should the scheme simply buy or sell all the animals herders want to trade, or should a pro-active attempt be made to monitor livestock numbers and the available fodder supplies and to seek a level of trade that keeps the two in rough balance (for example, Toulmin's "tracking strategy")?

Difficult questions also arise about where the purchased animals will be parked during the drought, and who will look after them and pay the costs. At least with feed subsidies, the herders continue to look after the animals themselves during the drought at no additional cost to the government. Also, if the program seeks to influence livestock prices through its market activities, what will stop animals' being brought into the region from elsewhere to take advantage of the higher purchase prices paid by the program during the drought, or of animals' being purchased at the program's lower selling prices in the post-drought period for transport to other regions?

Proponents of restocking programs claim a number of advantages, including the following:

- Restocking helps protect the capital assets of herders and enhances their food security.
- If the animals sold or given away are targeted to the more impoverished herders, restocking can help achieve a more equitable distribution of livestock and a more equitable use of common grazing areas.
- Restocking enables herders to more quickly build up their herds

after a drought, and to take maximum advantage of grazing resources as they recover.

- Restocking can help get herders away from settlement areas and relief-distribution points that are overcrowded and degraded to more remote areas that are underused, with positive environmental benefits.
- If herders know that they can restock quickly after a drought and at a reasonable price, they are less likely to carry too many animals into the drought and can adjust their herd-culling practices during the drought to enhance overall productivity.
- By providing the animals and not just the means to purchase them, restocking helps to avoid inflation in livestock prices in the immediate post-drought period. This contrasts with credit or other forms of financial assistance for post-drought recovery that merely give herders the financial resources to bid for the remaining stock of live animals in the market.

In recent years, restocking programs have been widely promoted by nongovernmental organizations and governments across Africa. Heffernan (1998) estimates that at least $100 million has been spent on restocking programs in Sub-Saharan Africa in the past decade, although this has not all been in response to drought. However, so far, little attempt has been made to rigorously monitor the impact of restocking programs on the productivity of pastoral systems, and economic analyses of their costs and benefits are not yet available. Anecdotal evidence suggests that livestock traders have been important beneficiaries from some restocking programs, and that any equity improvements in the distribution of livestock are short lived. There is also some evidence to show that impoverished herders have increased their incomes and consumption in the short term as a result of receiving livestock, but this should hardly be surprising and typically results from any direct-income or wealth-transfer program.

New Possibilities for Improved Drought Management

A limitation of most drought-management interventions is that they inadvertently subsidize inappropriate farming practices and encourage moral hazard. They also represent a fiscal burden to governments and donors that may be hard to sustain over the years. Two newly emerging approaches offer avoidance of these problems by providing farmers and herders with the means to better manage drought risks themselves, with minimal government intervention.

RAINFALL INSURANCE

Agricultural insurance has often appealed to policymakers as an instrument of choice for helping farmers and agricultural banks manage climate risks, such as drought. Indeed, many billions of dollars of public money is spent each year on agricultural insurance around the world. However, the experience has generally not been favorable. Publicly provided crop insurance has without exception depended on massive subsidies from government. Even then the performance of such insurance has been plagued by the moral hazard associated with many sources of yield loss: high administration costs, political interference (especially with compensation payments in election years), and the difficulties of maintaining the managerial and financial integrity of the insurer when government underwrites all losses.

Livestock insurance that compensates for loss of animals or reduced productivity because of drought has rarely been offered, and seemingly not at all for herders in traditional pastoral systems. There are good reasons for this: the incidence of drought losses is usually too high to make the insurance affordable; opportunities for fraud and moral hazard are too great; and there is little opportunity for on-farm inspection of management practices or loss assessments, particularly when the animals are on the move.

However, given the frequent occurrence of drought and the widespread damage that it causes, some form of insurance against drought losses is clearly needed. Indeed, if such insurance could be successfully designed, it might well displace the need for public drought-management policies. What is needed is a form of insurance that is affordable; is accessible to all kinds of people, including the poor; compensates for total income losses to protect consumption and debt repayment capacity; is practical to implement given the limited kinds of data available; and can be provided by the private sector without the need for government subsidies.

Area-based rainfall insurance offers a promising new alternative that in principle can meet all the requirements listed above. In this approach, rainfall insurance contracts are written against specific rainfall outcomes (for example, drought or flood) at a local weather station. The rainfall events should be defined at catastrophic levels, and they should be highly correlated with the value of regional agricultural production or income. For example, an insured event might be that rainfall during the most critical month of the growing season falls 70 percent below normal. In years when the insured event occurs, all the people who purchased the insurance receive the same payment per unit of insurance. In all other years, no payments are made.

Insurance is sold in standard units (for example, $10 or $100), with a standard contract for each unit purchased, called a Standard Unit Contract. Purchasers decide how many of these contracts to buy. The insurance is sold on a full-cost basis, and the price of the Standard Unit Contract is the premium. The insurance must be sold before season-specific information about the insured risk becomes available. This requires a purchasing deadline (such as a month before the normal arrival of the rainy season), after which new contracts are not sold.

Area-based rainfall insurance has a number of attractive features:

- It avoids all the moral hazard and adverse-selection problems that plague crop-insurance programs.
- It could be very inexpensive to administer.
- It uses only rainfall data, which are available in most countries for long periods of time.
- The insurance can be sold to anyone, including agricultural traders and processors, farm-input suppliers, banks, shopkeepers, and agricultural workers.
- It would be easy for the private sector to run.
- As long as the insurance is voluntary and unsubsidized, it will only be purchased when it is a less-expensive or more-effective alternative to existing risk-management strategies.
- A secondary market for insurance certificates could emerge that would enable people to cash in the tradable value of a Standard Unit Contract at any time.

In designing an area-based rainfall insurance scheme, a number of difficulties need to be overcome, including the following:

- The insurer faces high risk because of the covariate nature of the insured risk. When a payment is due, all those who have purchased insurance against the same weather station must be paid at the same time. Moreover, if the insured risks at different rainfall stations are highly correlated, the insurer faces the possibility of having to make huge payments in the same year. To hedge against this risk, the insurer can either diversify regionally by selecting weather stations and risks that are not highly (positively) correlated, or seek reinsurance in the international financial markets.
- Rainfall stations must be protected to prevent possible tampering of rainfall measurements. Possible approaches include more secure, tamper-proof stations and instruments.
- The actuarial soundness of the insurance could be undermined by El Niño weather cycles that change the probability of the insured events. It may be necessary to adjust the cost of the insurance whenever an El Niño event is confirmed.
- The volume of insurance sold could be too small to be profitable. The insurance will only appeal to people whose economic losses are highly correlated with the insured rainfall event. If the basis risk (the uninsured part of a person's risk) is high, the insurance will not sell. Also, if the probability of the insured risk is high, the cost of the insurance could be prohibitive. To overcome these problems, the insurance should be limited to truly catastrophic droughts that significantly affect agricultural production in a region.

The private sector might be expected to take the initiative in developing rainfall insurance, but several setup problems might require government intervention to jump-start activity in developing countries. These include paying the research costs of identifying key catastrophic rainfall events that correlate strongly with agricultural production and income, educating rural people about the value of rainfall insurance, ensuring secure rainfall stations, establishing an appropriate legal and regulatory framework for rainfall insurance, and underwriting the insurance in some way (perhaps through contingent loans) until a sufficient volume of business has been established that international reinsurers or banks are willing to come in and assume the underwriting role. These roles need not be costly but could prove crucial in launching rainfall insurance. However, it is also important not to launch the insurance on a subsidized basis, so as not to distort incentives for private insurers or farmers and herders.

EARLY-WARNING FORECASTS OF DROUGHT

In principle, the ability to provide early-warning forecasts of drought could be a powerful tool for avoiding many of the economic costs associated with the misallocation of resources that arise because farmers, herders, and other decision makers have to commit resources each year before key rainfall outcomes are known. For example, decisions about planting crops (including the date of planting, the seeding rate, and initial fertilizer treatment) often have to be made at the beginning of the wet season—before knowledge about rainfall outcomes is available. The economic value of season-specific forecasts really depends on the degree to which farmers can adjust their plans as the season's rainfall unfolds. If decisions about planting and cultivation practices—and the feeding, culling and seasonal movement of livestock—can be sequenced, with key decisions being postponed until key rainfall data are available, forecast information will be less valuable. However, if most decisions have to be made up front each season, the scope for mistakes will be much larger and the potential economic gains from reliable forecast information will be greater.

Stewart (1991) examines how the date of onset of the rainy season can provide a useful forecast of the ensuing seasonal rainfall pattern for Niamey, Niger, and shows how this information could be used to adjust planting and input decisions more optimally for the season (his "response" farming approach). Barbier and Hazell use a stochastic programming model to show how many of the decisions in a typical agropastoral community in Niger can be optimally adjusted to rainfall outcomes.

Reliable drought forecasts could also enable governments and relief agencies to position themselves each year for more effective and cost-efficient drought interventions. This possibility has already been realized, and several early-warning drought-systems are already in place in Africa that have proved successful in giving advance notice of emerging drought situations. However, these programs are really monitoring systems that track emerging rainfall patterns within a season rather than true weather-forecasting systems that predict rainfall outcomes before they even begin.

Reliable multiyear rainfall forecasts are not yet possible, but seasonal (from three to six months out) forecasts have become more reliable, particularly where an important part of the year-to-year variation in seasonal rainfall can be attributed to the Pacific El Niño Southern Oscillation weather patterns. As the ability to model these phenomena at the global and regional levels improves, it seems plausible to expect that more-reliable seasonal forecasts will be available at more local levels. This may prove to be one of the most exciting developments for drought management in the next few years. Private weather-forecasting services are likely to expand and become more available to developing countries. However, this is also an area where government could play a catalytic role, and even subsidize many of the development costs without having to worry that this would distort resource-management incentives at the farm level.

DROUGHT POLICY RESPONSES AND POLICY DEBATES

FEDERAL DROUGHT POLICY OVERVIEW

State governments can apply for areas to be declared in exceptional circumstances by the Federal Government. This then entitles farmers and agriculturally dependent businesses to access a range of economic support including exceptional circumstances Relief Payments through Centrelink. Receipt of payments is used as an eligibility criteria for some drought support programs (e.g. the Victorian Municipal Rates Subsidy)

The main federal drought support programs currently in place include:

- Interest rate subsidies;
- Exceptional circumstances Relief Payments for farmers and small businesses (through Centrelink);
- Exceptional circumstances Health Care Cards and associated concessions;
- Funding for emergency relief i.e. through the Country Women's Association emergency Drought Aid Fund;
- The Rural Financial Counselling Service Program and funding of additional counselling services;
- Grants for financial planning and advice e.g. FarmBis and Farm Help;
- Drought Force (Work for the Dole);
- Community strengthening programs such as 'Local Answers'; and
- 'Exit grants' to support farmers to leave farming.

Most of these programs are time limited and/or linked to exceptional circumstances status.

Victorian drought policy overview

Drought policy in Victoria falls largely within the operations of DPI, DSE and DHS.

Similar to the Federal Government, Victorian drought policy distinguishes between support for farm families and support for farm businesses.

The main Victorian drought support programs currently in place include:

- municipal rates subsidy;
- support for farm business to better manage drought conditions;
- rebates for low water allocations;
- rebates for water cartage and tanks;
- the Drought Apprenticeship Retention Bonus;
- mental health support; and
- social support initiatives.

In October 2007, the Victorian Government announced $100 million in additional funding for drought assistance to 'help rural communities battle the worst drought on record'. This funding included a range of targeted social and community support measures including:

- Additional mental health services;
- Emergency Volunteer Support Grants;
- Additional funding for drought employment programs;
- Funding for local employment and community strengthening programs; and
- Additional funding for drought coordinators.

Water supply security has been a major policy area for DSE with recent announcements of significant spending on infrastructure, including a desalination plant, pipelines to transfer water between regions and upgrades to irrigation systems. Our Water Our Future and associated plans form the framework of water policy in Victoria.

EMERGING DROUGHT POLICY ISSUES AND DEBATES IN VICTORIA

Strengthening and supporting rural communities

Community strengthening and social capacity building fit within the drought policy approach of supporting preparation, resilience and recovery. The Victorian Government has funded a number of programs in drought affected communities with these aims in mind. The Victorian Drought Social Recovery Strategy, coordinated by the DHS Emergency Response Unit, provided funding to employ local community development officers to support community strengthening and drought recovery activities including social gatherings. The aim of this program was to build community adaptive capacity using local resources.

The Rural Women, Drought and Climate Change Initiative under the DPCD is a two year project (which commenced in 2007) that employs five local coordinators to develop networks and strategies to support rural women's role in supporting adaptation to climate change and drought.

In October 2007, the Victorian Government announced funding for drought coordinator positions within local councils in exceptional circumstances declared areas. This program will aim to provide coordination and brokerage of drought responses at the local level.

While there is a clear community strengthening and resilience focus of these projects, the efficacy of these community strengthening approaches may have been undermined by short time funding models. For example, the Victorian Drought Social Recovery Strategy was designed as a four year project, and was funded in four separate nine month funding rounds. Programs funded by DHS such as the 'Sustaining Community Wellbeing' program and the 'Tackling Mental Health Initiative' were targeted at specific areas deemed to be severely drought affected and also received only time limited funding.

Short term funding creates difficulties in staff recruitment and retention and leaves programs vulnerable to disruption due to changes in staff – a significant issue in work which is centred around networking. Even where funding for programs is extended, uncertainty leading up to funding announcements contributes to staff turnover. These short term funding models reflect an 'emergency response' approach to drought policy.

Many programs are also focussed on farmers and may not address the needs of other low income and disadvantaged groups within drought affected communities. Community strengthening programs need to be more widely and sustainably funded and better integrated with other social and economic transition activities to develop 'bridging capital' and to function more effectively to increase community resilience.

Education and training

The economic stresses of drought impact on the education of rural young people. Several policy initiatives at the Victorian and federal level have been developed to address these issues. The federal Drought Force program, which allows people in drought affected areas to undertake farm work as part of a Work for the Dole scheme, provides benefits to rural young people by allowing them to work on the family farm while accessing a NewStart allowance.

Some funding for skill development is available through the Federal Government to support drought affected farmers to either gain accreditation for additional skills, such as trades, or to retrain on exiting farming.

The Victorian Government has put in place a number of drought policies related to education assistance including:

- 'Back to School' funding, administered through VicRelief and Foodbank, which distributes funding to meet school costs through local community agencies;

- assistance for families in drought affected areas who are struggling to pay kindergarten fees; and
- apprenticeship retention support.

Evaluation of the effectiveness of these programs and the distribution of access to them within drought affected community in Victoria would greatly assist in future policy making around education and training.

The mental health impacts of drought are significant and widely recognised. In October 2007, the Victorian Government announced an additional $587,000 funding for mental health services in drought affected areas. This investment builds on a previous funding which was due to expire in December 2007. Appropriate service delivery models for drought related support services – such as outreach models which support anonymity of service users in rural communities - are essential.

Short term funding models for drought social support services impact on the retention of a skilled workforce and increases the cost of program delivery. The financial and social effects of drought and their health implications adds greater urgency to issues associated with limited access to bulk billing health services in rural areas and recruitment difficulties for rural health and mental health professionals.

ROLE FOR VCOSS

Climate change, drought and equity cross sector advocacy

Most coordinated advocacy in the area of climate change and equity within the non-government sector has been focussed at the federal level, particularly in the context of a federal election year. For example, the Brotherhood of St Laurence (BSL), the National Welfare Rights Council, the Australian Conservation Foundation (ACF) and the Climate Institute convened a 'Climate Change and Equity' roundtable in March 2007. The BSL's policy and research unit is developing a research program in this area and is in the process of employing research staff. Australian Council of Social Service, the ACF and Choice have also formed a partnership to produce a report on energy efficiency and equity.

There is a gap in such coordinated advocacy at the Victorian state level - given that state and local governments are likely to remain responsible for many policy areas particularly those around adaptation, in which social and equity impacts of climate change are likely to be an issue. VCOSS could have a role in developing networks between the environment and social justice sectors to advocate at the state level.

In the UK, the community sector has developed the 'Third Sector Declaration on Climate Change' as a public statement of the link between social justice and climate change issues. The declaration summarises the imperative for the community sector to address environmental issues as organisations and advocates. The adoption of a clear policy position which identifies climate change as a social justice issue is a possible first step towards coordinated advocacy across the non-government sector.

VCOSS could provide leadership in this area through the development and endorsement of a similar statement.

Proposed Priority Policy Areas for VCOSS

The social justice and equity implications of climate change, drought and policies to address these issues create a case for VCOSS intervention in a number of policy areas. In some cases these issues are already addressed within existing VCOSS policy areas. For example, climate change issues are already incorporated as a secondary consideration into policy work around transport and utilities. However, including climate change as a social justice issue in and of itself has not been part of VCOSS policy work to date.

Some work around the impact of drought on service provision in the context of climate change is already undertaken by VCOSS in the areas of emergency relief and health, however drought and accompanying rural social and economic adjustment have not been actively considered as a long term drivers of social policy.

A clearer mandate to work directly on equity and climate change issues would increase capacity, expertise and visibility in specific aspects of policy related to climate change and drought. This would allow VCOSS to engage more actively in policy development and critique in these areas.

The following policy areas are considered the most immediate and strategic for VCOSS to begin actively incorporating into policy work:

- energy and water pricing and household efficiency;
- heat waves and health inequalities;
- responses to increased food costs;
- domestic water security;
- social and economic support policies associated with drought; and
- housing standards and insurance.

Supporting Inclusive and Equitable Climate Change Policy and Research

It is vital that the community sector engages with policy and research developments to encourage and facilitate the inclusion of equity considerations and the voices of disadvantaged groups. Research and policy development around climate change within universities and government is broadening from looking at purely eco-systemic and environmental impacts to looking at social impacts and adaptation. Ensuring that there are adequate baseline measures and appropriate indicators which can be used to monitor the impacts of climate change and climate change policy on different vulnerable groups is important. Similarly there is a role for the community sector in assisting researchers and policy makers in including direct information about impacts on vulnerable communities in their work. At the local level this involvement is also critical. Regional work into understanding the impacts of climate change on communities and assessments of adaptive capacity, for example, need to reflect diversity within communities. The community sector can have a role in ensuring that disadvantaged communities are able to participate in consultation and planning processes as well as representing these groups and advocating on their behalf.

VCOSS's role in undertaking social research and links to the research community and government provide an important opportunity to proactively ensure that social and equity considerations are incorporated from an early stage. This is essential not only to avoid inequitable outcomes, but also to avoid social and equity concerns being manipulated as a justification to avoid action.

Impact of Drought and Climate Change on Community Service Delivery

Climate change and drought will be increasingly prominent in the operating environment of community sector organisations. VCOSS could take the lead in documenting the impact of climate change and drought on the community sector and advocating for appropriate and timely strategic planning in this area.

The community sector delivers a range of services which may face increasing demand under the effects of climate change and drought. These include; health and mental health services, support for older people, emergency response and emergency relief services. It is important that services are able to be expanded to meet this increased demand. The ability of the community sector to respond to changing community needs will require appropriate and timely service planning and adequate funding. The community sector may need to continue to advocate for adequately funding for these services supported by evidence of changing service demand and unmet need. Outreach models suitable for drought support work in rural areas may raise occupational health and safety issues for unaccompanied workers 'cold calling' clients who may be in crisis in areas with poor mobile phone coverage. The cost of outreach service delivery models over large service areas is also problematic in the context of high petrol prices.

Community based organisations have strong links into local communities, especially into groups within communities who may have difficulty accessing information or may be hard to reach with mainstream marketing and communication strategies. The community sector's role as a conduit of information to vulnerable and disadvantaged groups will be vital in ensuring that information about climate risks is distributed to, and understood by, those who need it most. For example, information about preparing for heat waves or other emergencies, opportunities for climate proofing or increasing water and energy efficiency within the home and support services or material assistance available during droughts, fires or other emergencies will reach vulnerable groups more effectively through the utilisation of community sector networks.

Supporting Community Sector Sustainability

The community sector, like all other parts of the Australian community, has a responsibility to take action to prevent catastrophic climate change, especially given the equity implications at a global level. There is growing interest within the sector in incorporating environmentally sustainable organisational practices. Community sector organisations will also need to adapt to changing conditions under climate change including higher utilities and fuel prices as part of broader sector sustainability planning. The community sector has scope to reduce their environmental impact and vulnerability to price rises but may benefit from education, support and funding incentives to do so given limited financial resources.

There are numerous programs, networks, information services and grant opportunities to support the business sector to reduce their environmental impact, some of which may be available to the community sector.

The community sector may also choose to become involved in the delivery of environmental programs, especially those like household retrofitting which will directly benefit low income groups.

In response to these emerging issues, opportunities for VCOSS' in sector development include work to support and encourage the sector to function more sustainably. Work in this area could include:

- information distribution;
- coordinating training and professional development;
- advocating for increased access to grants programs;
- brokering of partnerships and access to sustainability services;
- development of environmental management strategies for the community sector; and
- investigation of the use of ICT to reduce fuel consumption.

7

The Impacts of Water Management During Drought

Many rivers have been degraded by loss of habitat, changes in flow regime and poor water quality. These impacts can exacerbate the stresses already experienced by aquatic fauna and flora during drought periods. Where ecosystems are already degraded, the ability to survive may be damaged, and the speed of recovery inhibited. For instance, high erosion rates can cause a massive inflow of sand and silt into the main river channel, and this can greatly reduce the availability of deep waterholes as drought refuges. Similarly, the extraction of water from the remaining waterholes along a river channel, can greatly diminish the resilience of the river's biota to drought.

A high demand for water during drought often results in a large drawdown of lakes and reservoirs. As the nutrient-rich bottom waters move closer to the surface, the increased availability of light can result in higher rates of algal growth, and in some instances, the formation of blue-green algal blooms. During drought periods, low flows through weir pools can result in stratification of the water, and this also increases the chance of algal blooms. The inflow of polluted water from point sources, with little subsequent dilution, can also contribute to poor water quality conditions in rivers during low flow conditions.

Problems such as eutrophication, the formation of algal blooms and the development of low oxygen conditions, all increase the stress already experienced by river biota during drought. For example, the low oxygen conditions that result from the growth, death and decomposition of large algal blooms in weir pools or reservoirs, can result in severe fish kills.

Along regulated rivers, the floodplain vegetation is already stressed by the reduced frequency of flood inundation. During a drought this stress is heightened by the extended period of drying, and also from elevated concentrations of groundwater salinity. Each of these issues needs to be carefully addressed by river managers to ensure that the impacts of drought on regulated rivers do not result in permanent loss of biodiversity or a long-term reduction in river health. In severely degraded rivers and wetlands, a prolonged drought may be 'the straw that breaks the camel's back' for some of the sensitive or threatened species.

BETTER MANAGEMENT OF RIVERS DURING DROUGHT

The way we manage the land and water within a catchment, both during and following a drought, is of critical importance to the health of our rivers. In particular, in-stream and floodplain refuges, and the organisms they support, need to be protected. Some management actions that can be taken to protect aquatic communities are discussed below.

Large dams greatly alter the natural flow regime of rivers. If dams are emptied during drought, it can take months or years until they refill and spill again. In such cases, the section of river immediately downstream of the dam may experience a greatly extended drought period. Periodic flow releases, including base flows and pulses, can help the downstream aquatic communities survive and recover from these droughts. These releases might also reduce the risk of stratification in downstream pools, which in turn decreases the chance of an algal bloom. It might also reduce the risk of deoxygenation or poor water quality in stagnant pools, which can pose a risk to the native fish populations. Although some may consider these flow releases as a waste of precious water during the drought, it is a small and very sound investment in sustaining healthy rivers.

During drought conditions, native fish tend to congregate in the remaining deep waterholes along a river channel and are easy targets for fishing. If a large number of the remaining fish are caught, their rate of recovery after the drought will be greatly inhibited. This is also true in large water storages, where the remaining fish might have to congregate in only 5 or 10 per cent of the maximum storage volume. During these times, it would be wise to introduce restrictions on fishing, to ensure that sufficient adult fish survive the drought for future breeding.

When river flow ceases and the channel bed starts to dry up, the remaining waterholes become refuges for birds, fish, amphibians and invertebrates (such as yabbies and insects). It is critical that these waterholes are not pumped dry when providing water for livestock. Alternative sources of water should be considered where possible. During droughts, livestock tend to congregate along riverbanks and around the remaining waterholes. Overgrazing, trampling and high levels of dung can add severely to the stress that these habitats are already experiencing. Stock should be excluded from riparian lands and alternative watering points should be provided away from the river.

Managers of drinking water supplies also need to be aware that drought-breaking storm events are likely to bring significant 'dirty water' into rivers. Storm run-off will pick up high concentrations of fine sediment, nutrients and pathogens from the dry soil surface, especially during the 'first flush' of heavy rains. In south-eastern NSW, northern Victoria and the ACT, the drought of

2002/03 resulted in widespread forest fires throughout the entire Snowy Mountains region. The loss of protective vegetation cover greatly increased the risk of poor water quality during the first rainfall events after the fires, leading to high loads of sediment and ash being washed downstream. In the ACT, where 90 per cent of the Cotter River water supply catchment, was burnt, floating booms and curtains were installed at strategic points in the dams and reservoirs to strain out ash and other debris that could otherwise enter the water supply. If water quality is affected by such events, it may be prudent to bypass the first flush of dirty water or revert to alternative water supplies.

Increasing our knowledge of droughts

To develop best practices for river management during drought, we need a comprehensive understanding of how freshwater aquatic systems respond to reduced (or no) flow. However, our knowledge is still very limited, and the Cooperative Research Centre (CRC) for Freshwater Ecology is currently trying to address this shortfall.

The 'Dryland River Refugia' project is investigating rivers in the northern regions of the Murray-Darling Basin and in the Lake Eyre Basin where episodic floods are often followed by extended periods of drought. For much of the time these rivers exist as a network of turbid waterholes along ephemeral channels.

This environment provides an ideal opportunity to determine the importance of waterholes as refuges for aquatic organisms. The project is investigating the relationship between biodiversity and the physical attributes of individual waterholes, and how populations are connected in space and time. It is also identifying the biophysical processes that sustain biodiversity and ecosystem health within these aquatic ecosystems. The 'Granite Creeks' project in northern Victoria, is investigating how aquatic organisms disperse, and how far they can move. This provides valuable information on the ability of plants and animals to recolonise sites after a disturbance such as drought.

The 'Cotter River' project, in the ACT, is investigating how to maximize the ecological benefits of environmental flow releases below two large water storages. These releases should help maintain suitable habitat for native fish species, particularly during extended droughts when the impacts of the dams would otherwise result in unnaturally long periods of little or no flow.

These, and other research projects by the CRC for Freshwater Ecology, will help us to increase our knowledge base, and should lead to better management of rivers and their catchments during periods of drought.

IMPROVING RIVER MANAGEMENT DURING DROUGHT

WHAT IS A DROUGHT?

Drought can be defined as an unusually long period without rain. More specifically, the Australian Bureau of Meteorology identifies drought under two rainfall deficiency categories:

- A severe rainfall deficiency exists in a district when rainfall for three months or more is in the lowest 5 per cent of records for that area.
- A serious deficiency lies in the next lowest 5 per cent (i.e. lowest 5 per cent to 10 per cent) of historical records for three months or longer for that area.

The wide range of climatic conditions that exist in Australia means that a long period with little or no rainfall might constitute a drought in one region but not in another. For instance, many parts of northern Australia receive little or no rainfall for 3-4 months in the middle of each year, but this is not a drought since it is a predictable event that occurs every year. However, in other regions that have high rainfall all year round, a period of 3-4 months without rain will cause a severe drought.

DROUGHTS AND ECOSYSTEM DISTURBANCE

Extreme events, such as droughts and floods, play an important role in the functioning of our river systems and in the maintenance of biodiversity. A drought can be considered as an ecosystem disturbance (i.e. reduced water availability) to which plants and animals respond. The response of biota can take two forms: resistance is the capacity of the biota to withstand the drought, while resilience is the capacity of biota to recover from the drought.

In some regions, relatively predictable dry seasons occur each year, and freshwater biota have evolved adaptations, such as life history scheduling, or the use of refuges such as deep waterholes, to survive them. In general, the response of biota to predictable dry seasons is characterised by a high resistance and a strong resilience. Droughts, however, are less predictable and this makes it more difficult for organisms to develop adaptations and strategies for survival. The response of biota to droughts is characterised by a low to moderate resistance and a variable resilience, making recovery more difficult or requiring more time than for seasonal dry periods.

The predictable nature of seasonal events has tended to make them easier to include in ecological studies. Less is known about droughts and much of our knowledge has been derived from opportunistic studies that were initially set up for other purposes.

What Happens to River Systems as they Dry?

A drought is characterized by a sequential decline in rainfall, surface runoff, soil moisture, groundwater interaction, and ultimately discharge, for naturally flowing rivers. As the drought progresses, floodplain and wetland areas dry, and rivers become confined to their low-flow channels. These dynamic processes play a key role in the long-term maintenance of freshwater biodiversity. Although there is a widespread reduction in aquatic habitat, there is a corresponding increase in the area of semi-aquatic and terrestrial habitat. For instance, as the water recedes, new plants germinate in the exposed mud and waterbirds such as ibis and spoonbills feed in the shallow pools and mudflats.

As river flow ceases, fish and other aquatic organisms seek refuge in the remaining pools, where the interaction of predation and competition can markedly alter community composition. Many terrestrial animals also rely heavily on these remaining waterholes for their drinking water. Any macroinvertebrates or fish that become stranded in dried up habitats, provide an abundant food source for terrestrial predators and scavengers, such as birds and ants.

As waterholes decrease in size, the death and decomposition of aquatic plants and animals releases nutrients such as nitrogen and phosphorus into the remaining water. Evaporation slowly concentrates these nutrients even further. High nutrient levels can result in dense growths of algae, particularly during summer. These are grazed by macroinvertebrates, which in turn are eaten by fish. The algae may create large diurnal changes in oxygen concentration, and if severe algal blooms occur, the low overnight oxygen levels might ultimately threaten aquatic biota, such as fish.

Evaporation also causes an increase in salinity, and this affects salt-sensitive plants and animals. The intrusion of saline groundwater can also increase as the water level in the river channel drops. Groundwater intrusion can cause stratification in pools, leading to further deterioration of water quality and a higher chance of algal blooms.

Resistance Mechanisms for Drought

Organisms resist drought using a range of physiological, behavioural and life stage responses. For instance:

- Fish move downstream or retreat to isolated pools as headwaters dry.
- Plants reduce evapotranspiration losses from leaves and set seed.
- Yabbies, some frog species, and insects such as water beetles, survive by burrowing deep into moist sediments. These animals can remain safely buried for many months or even years.
- Turtles migrate overland to find new waterholes, while other species can hibernate for at least 3 months without water.
- Zooplankton, some benthic invertebrates, algae and bacteria produce desiccated, drought resistant spores.
- Waterbirds fly many hundreds of kilometres to find more permanent water, such as coastal estuaries.

In spite of these resistance mechanisms, some populations may suffer large losses during a drought and take many years to recover. However, the life cycles of other species are highly dependent on an extended drying phase, and their long-term survival is dependent on a cycle of floods and droughts. Also, some introduced pests have poorly developed resistance mechanisms for drought, and this can provide a competitive advantage to drought tolerant native species.

What happens when the drought breaks?

Australia's freshwater flora and fauna are generally well adapted to patterns of flood and drought and have developed mechanisms for recovery. However, the rate of recovery can vary considerably, being influenced by succession patterns and whether or not any species that have been lost can be restored. Recolonisation by fauna from drought refuges is important in this recovery. A problem arises if refuges such as permanent waterholes or billabongs have been eliminated by river regulation, water extraction or farm development. Any loss of previously abundant species can result in major changes to community structure and a lag in recovery.

When the drought breaks, the initial flush of water downstream is likely to bring large quantities of sediment and organic matter that has steadily accumulated in the catchment during the drought. The organic matter and dissolved nutrients provide an abundant food supply and this can result in a mass hatching of invertebrates. This in turn provides food for higher organisms such as insects and fish. Highly mobile species, such as waterbirds, quickly move in from surrounding regions to take advantage of these new food supplies. The community that existed before the drought may not fully re-appear until there has been a complete progression from early colonizers through to species with longer life cycles. Some species, such as native fish may take many years to fully recover but, over decadal scales, there is little evidence that fish populations are affected by drought in healthy river systems.

CONSECUTIVE DROUGHTS OF RAJASTHAN THROUGH PARTICIPATORY WATER RESOURCE DEVELOPMENT

The droughts of Rajasthan are well known. Erratic and low rainfall and extreme temperatures are the common features. The yield of food grains is just 803 kg per hectare and is the lowest in India. The all India average food grain yield is 1502 kg/ha Percent population below poverty line (BPL) in Rajasthan ranges from 30 to 46 per cent in different regions. Overall literacy rate of Rajasthan is (2001) 61 per cent, of which 76.5 per cent is in male and 44.3 per cent in female.

Most of the regions in Rajasthan fall under Hot Semi-arid zone. Out of total culturable wasteland of 13.879 million ha in India, as of 1997-98 the state of Rajasthan has a share of 5.017 million ha being highest in the country. Rajasthan has a common property land area of 35 per cent. The National Sample Survey (NSS) in its 54th round (1998) has estimated that common property lands constitute 15 per cent of the geographical area of country and that 45 per cent of all rural households collect fuel wood from common property lands. (ITCOT) The villages Gokulpura, Govardhanpura, Thana and 4-5 hamlets in Bundi district of Rajasthan were not different from the villages in the region. But now they have gone far ahead of their neighbors. There has been no effect of successive droughts of Rajasthan on these two villages. They have sufficient water at their door-steps for drinking and also harvested abundant crops in both Kharif and Rabi. This is due to the phenomenal impact of a development programme on watershed basis implemented during 1996-2002 by BAIF, an NGO based in Pune with financial assistance of India Canada Environment Facility (ICEF), New Delhi. Following sections of paper describes the programme strategies, activities and outcomes.

Project Area

Project area covers the villages Gokulpura, Govardhanpura, Thana and 4-5 hamlets in Bundi district of Rajasthan. Total geographical area of watersheds is 3286 ha and is owned by 633 families and two gram-panchayats. Project has been implemented since 1996-97.

The major portion of the population is scheduled tribes and scheduled castes. The baseline survey of 1997 revealed very poor literacy rates. Just 12.7 per cent of women had gone to primary school, 3.01 per cent up to secondary school and 84.2 per cent were illiterate. In case of male, 36.02 per cent had gone to primary school, 24.4 per cent up to secondary school and 36.3 per cent were illiterate. About 80 per cent households were consuming the total grains produced and had no surplus to sell. The dependency on moneylenders was very common in the area. Livestock owned by villagers was unproductive due to severe genetic erosion and fodder scarcity. Social customs like purdah, denial of girl's education, teenage marriages and male dominance have deep roots in the society preventing women to occupy common platform for development. Women of this Meena habitat were the most sufferers.

Rainfall failure is common in the project area. Further, the rainfall is erratic. For instance, in 1998 the watershed received only 245 mm. In 1999 the rainfall was 438 mm; in 2000 it was 399 mm. In 2001 the watershed received 601 mm, which included 344 mm that fell over just three days. This kind of rain causes more harm than good, as it washes away a lot of soil. During the four rainy months of 2002, the area received just 200 mm of rain.

Undulating terrain, low soil fertility, denuded wastelands with rock outcrops (both common and individual) and high rate of soil erosion, were the typical features of area. The area lies in the West of Great Boundary Fault with an entire metamorphic terrain. The sedimentary rocks, which have been metamorphosed, are characterized by a pervasive foliation whereas the metavolcanics do not display such strong foliation. The phyllite, metavolcanics, schists and sedimentary rocks form the major lithology of the area. The Aravalli sub-group rocks are also exposed in a small portion.

Project Measures

The key approach was development of common and individually owned wastelands, adoption of traditional and modern techniques for water recharge and improved irrigation practices with management and sharing of water by the people themselves. In addition to these core measures, energy conservation, drinking water resource development, livestock development, training and awareness generation among the community were also part of the development. With initial resistance, reluctance and arguments, slowly people started coming forward after seeing the positive effect of demonstration activities (entry point activities) and selfless work culture of BAIF. It took almost a year to build a rapport with the community and motivate them for their active participation in the programme.

The activities were implemented through People's Institutions developed in the villages. The main institutions include 9 Village Watershed Committees (VWC), 2 Charagah Samittees (i.e. Silvipasture Management Committee), 13 Women's Self Help Groups (SHG) and a Men's SHG. The work done include Field Bunding on 637 ha, Silvipasture of 80 ha on Panchayat land and 16 ha on privately owned lands, Afforestation on 75 ha and Horticulture on 50 ha.

Under the Drainage Line Treatment, 1500 Gully Plugs, 36 Stone Bunds, 21 Permanent Checkdams/ Anicuts, 6 Underground Bandharas (Sub-surface dyke) and 13 Gabion structures were constructed. The indigenous knowledge of the people and scientific methods of geo-hydrological and engineering investigations blended well for initial assessment, planning and implementation of project activities.

The traditional water conservation measures were defunct. Almost all the bandhas, constructed during the period of Maharajas were not in use due to the want of major repairs. People were confident of high utility of these structures. Geohydrological and engineering investigations also indicated the high potential of ground water improvement if these structures were repaired. When repaired during 1999 to 2001, the structures proved to be the best fit in the region.

The structures were repaired along with treatments in their catchments. These structure and similar types of newly constructed ones changed the scenario of irrigation and drinking water. The fertile soil has been conserved in the storage areas of structures, which helped cultivate the rabi crop in the storage area itself. The ground water regime has drastically improved. The wells in the area of influence of the structure remained full during kharif season and had sufficient water for irrigation to rabi crops. The summer crops were also started coming up due to availability of water. The overflowing water was used for cultivating paddy. In the year 2000, which happened to be the drought year (rainfall of 399mm), first time people cultivated basmati rice.

A traditional water harvesting structure revived in the village Thana

Innovative measures like subsurface barriers using the clay soil in the drainage, gabion structures and an eco-friendly and low cost structure such as gabion with the locally made net using the grass, have proved very successful and effective structures of water harvesting.

Implementation Approach

All the treatment measures and water harvesting structures have been constructed by the locals after technical trainings and guidance from BAIF staff. The programme was implemented through the watershed committees, which are operating on the basis of set rules and regulations prepared jointly by the community and BAIF. The maintenance, management and sharing of benefits of common assets is being looked after by the Watershed Committees and User Groups. The children in the village are also actively involved in cleaning the drinking water sources (hand pumps and open wells), plantation of the trees in and around the school campus and keeping hygiene in the villages. The school children as well as the teachers were trained in environmental management and conservation and now they are actively involved in the programme with their "Nature Clubs" formed in each village and hamlet.

The Resource Teams have been created from the active community members. These teams are the main link for sharing project experiences with Govt., NGOs, bankers and communities at various levels, i.e. block level to state level. See chart below for the Institutional arrangements in the project.

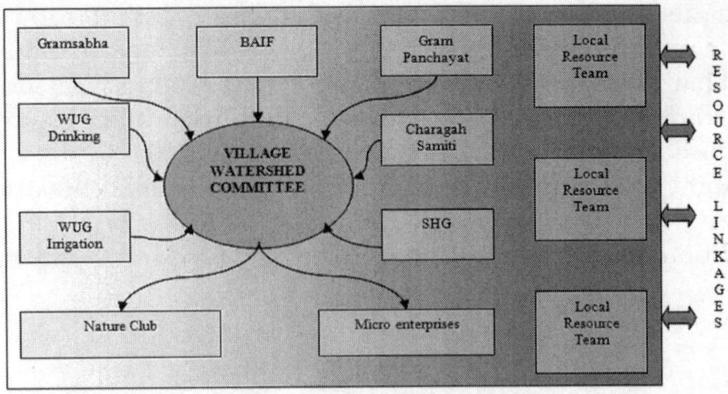

Fig. Chart: Institutional Mechanism

The specialty of the programme approach is distinct. It includes (a) a participatory approach in development; (b) training of the local youth in project activities; (c) use of advanced planning and designing facilities like Remote Sensing Data and GIS; (c) expert inputs in Engineering, Geo-hydrology, Agriculture, Cattle Development, Social Science; (d) close project monitoring; (e) blending of family based and area based approach in development and (f) close link between BAIF field staff and the villagers.

Project Results

Within the natural and social framework of adverse situation, three villages have come out of the evils. A true drought-proofing model has been emerged in the area. People of the area have been experiencing the effects of pioneering work of drought resistance during last three years. The constraints have turned into the productive assets. The common wastelands have turned into grasslands and forestry. The soil has been conserved. Rainwater recharges ground in contrast to earlier flash floods.

Drought Proofing

Wells remain full of water in kharif and rabi seasons. Beyond the imagination of BAIF, people have gone to cultivate the basmati rice. The surplus water flowing in the streams and water from some of the overflowing wells have been utilized for cultivation of basmati rice by 60 families. The area producing about ₹ 3000/- to 4000/- worth per acre, first time produced over ₹ 25,000 worth per acre in the year 2000 -01.

In the years 2000 and also 2001, this particular region was declared as drought affected area. Villagers of project area say, "To our surprise, the Government eliminated our villages from the list of drought affected villages. Of course we are very proud of the fact." The rainfall of 2000 was just 399mm. A degraded common wasteland has been converted into a productive pastureland. People have been managing it through a committee. Part of the grass is cut and taken by the farmers for their own cattle and the surplus grass is sold. So far the net profit from the silvipasture has gone up to ₹.82,876/-. The area under assured irrigation is now 84 per cent of the crop area of the watershed. Before watershed project, it was just 45 per cent protective irrigation and that too if the rains are good in that year.

The yield of the wheat, which was 2.7 tons per hectare, has now increased to 3.7 tons per hectare. Well above the national average of 2.5 tons per hectare and quite close to the top producing state average (Punjab 4 tons per hectare). Of course here the emphasis is on organic fertilizer application while in Punjab the irreversible process of land degradation has been happening due to over utilization of chemical fertilizers and water. Maize production has gone up from 1 tons to 1.9 tons per ha.

The Indian Remote Sensing satellite images show that the green cover in the area was below 10 per cent. Now in it ha gone up to 20 to 30 per cent.

Costs and Benefits

Total project investment till the end of 2002 was ₹ 200 lakhs. This has been invested over an area of 3286 ha, so the cost per hectare comes to about ₹.6000. People have contributed additional ₹ 1000 per hectare. The average of past two years shows that the additional annual income over and above the past income is ₹ 77 lakhs. A major sustainable source of income is expected from the newly grown horticulture. Table shows the increased income from agriculture in the villages Gokulpura and Govardhanpura. The income in the village Thana has been also increased substantially. The other income sources include livestock, micro-enterprises, and local agro-based employment opportunities.

Table. Increased Agricultural Income in Gokulpura and Govardhanpura

Crops	Total Production (Qt)		Total Income (Rs. lakhs)	
	1997	2000	1997	2000
Maize	1963	3330	8.83	18.32
Wheat	5719	7837	34.31	53.53
Other Crops			22.55	47.49
Grass	86.24	295	0.09	0.90
Total			65.78	120.24

Reduction of Women's Drudgery

Development of new sources of drinking water, revival of the existing defunct infrastructure and rejuvenation of defunct water sources helped solve the problem of drinking water.

Programme Sustainability

The project outcomes indicate a drastic shift from under employment, food and water scarcity to year round self-employment, food security and sufficient water supply for drinking and irrigation. The capacities of the communities have been built to manage resources developed and to make the development sustainable. People's institutions and trained locals were important steps of project. Till the end of September 2002, about ₹ 2.4 lakhs were raised by the committees as a common fund for future maintenance and management. This was also used as corpus for providing loans to the villagers and also mass procurement of fertilizers and seeds by the VWCs. The mass procurement has helped cut down the prices and transportation costs by the individuals. The common fund is under circulation and hence the amount will rise up.

IMPACT OF CLIMATE CHANGE ON WATER RESOURCES

The impacts of climate change on water resources have been highlighted in the Fourth Assessment Report of the Intergovernmental Panel on Climate Change (IPCC), which predicts an intensification of the global hydrological cycle, affecting both the ground and surface water supply. The IPCC report also concluded that it is highly likely "The negative impacts of climate change on freshwater systems outweigh its benefits", with runoff declining in most streams and rivers. In addition, different catchments are likely to respond differently to the same change in climate drivers, depending largely on catchment physiogeographical and hydrogeological characteristics and the amount of lake or groundwater storage in the catchment. The IPCC has predicted with high confidence that the drought affected areas will show increase in frequency as well as the severity of drought. The IPCC also predicts with high confidence that the area affected by drought will increase in South Asia, including India. The impacts of climate change are also dependent on the baseline condition of the water supply system and the ability of water resource managers, to respond to climate change in addition to pressures of increasing demand due to population growth, technology, and economic, social and legislative conditions.

SURFACE WATER RESOURCES IN KARNATAKA STATE

There are seven river systems in the state namely Krishna, Cauvery, Godavari, North Pennar, South Pennar and Palar as shown in Figure. The annual average yield in the seven river basins is estimated to be 98,406 Million Cubic Meters (MCM). However economically utilizable water for irrigation is estimated at 1,695 Thousand Million Cubic feet (TMC). According to the Karnataka State Environment Report (2003), there are 36,679 tanks in the state with an irrigation potential of 6,84,518 ha (0.6 Mha) and minor irrigation surface tanks with estimated irrigation potential at about 1.0 Mha.

Objectives

Studies so far on the impacts of climate change on river basins are limited to national level assessments. In this report, we concentrate on the state-level (Karnataka) assessment: the two important river basins of the state, Krishna and Cauvery, were selected for the present study. This study aims to assess the impact of climate change on water resources on these two river basins using the HadCM3 Global Circulation Model (GCM). This study will address:

- Precipitation in the two river basins
- Runoff or water yield
- Evapotranspiration
- Drought

River Basins Selected

Krishna and Cauvery river basins have been selected for the study and in this report we present a synthesis of available information on climate change impacts for the two river basins. These two are the major basins in the state accounting for nearly 78 per cent (Krishna- 59.60 per cent, Cauvery- 18.34 per cent) of the total basin area of Karnataka. Table: River systems of Karnataka, their average yield, area drained and percentage of water yield (Source: Department of Water Resources, Karnataka)

Krishna basin

Krishna river basin is the fourth largest in India in terms of annual discharge and fifth largest in terms of surface area, covering parts of three states in South India. It covers approximately 59 per cent of the geographic area of Karnataka.

The principal tributaries of Krishna in Karnataka are Ghataprabha, Malaprabha, Bhima and Tungabhadra. The main issue facing the river basin is that the construction of large dams on the river for generation of hydroelectricity has severely reduced the river discharge. In a study done by Bouwer et al. (2006) it was observed that climate variability varied the runoff by 6-15 per cent while the construction of reservoir on the river decreased the runoff by 61 per cent.

Table. Area under the Two River Basins

	Area (sq km)	% area of Karnataka under river basin
Karnataka	1,91,791	100.00
Krishna basin	1,14,312	59.60
Cauvery basin	35,168	18.34

Cauvery basin

The Cauvery river basin is the second major river system in Karnataka. It covers approximately 18 per cent of the geographic area of Karnataka. The principal tributaries of Cauvery in Karnataka are the Harangi, the Hemavathy, the Lakshmanathirtha, the Kabini, the Shimsha, the Arkavathi and the Suvarnavathy. All these rivers, except the Kabini, Arkavathy and Suvarnavathy rivers, rise and flow fully in Karnataka.

Methodology

An assessment of the impacts of climate change on water resources can be best handled through modeling of hydrological conditions in river basins under future predicted climate variables. The main components of the hydrological cycle are precipitation, evaporation and transpiration. Changes in climate parameters, solar radiation, wind, temperature, humidity and cloudiness affect evaporation and transpiration. The river basin is generally the most appropriate primary exposure unit for assessing impacts on hydrological resources. For the present study, three variables i.e. precipitation, evapotranspiration and surface runoff have been considered.

- Precipitation is any product of the condensation of atmospheric water vapor that falls under gravity. It manifests mostly as rainfall in a region.
- Evapotranspiration (ET) is a term used to describe the sum of evaporation and plant transpiration from the Earth's land surface to atmosphere. Evaporation accounts for the movement of water to the air from sources such as the soil, canopies, and water bodies. Transpiration accounts for the movement of water within a plant and the subsequent loss of water as vapor through stomata in its leaves.
- Surface runoff is the water flow that occurs when soil is infiltrated to full capacity and excess water from rain, melt water, or other sources flows over the land.

Global Climate Model (HADCM3): The process based, Global Climate Model, HadCM3, is used in this analysis for generating the A1B scenario for the three variables, precipitation, evapotranspiration and runoff. This model has been used recently for generating climate change projections for various parts of the Indian subcontinent (Rupa Kumar et al., 2006). According to this study, the Indian subcontinent could be subjected to an average of over 4 °C increase in temperature by 2085 under the SRES A2 scenario.

Data: Spatial data and the source of data used for the study include:

- Current observed temperature and precipitation: The 30-year climatological data from CRU
- Climate change projections: Regional Climate Model (RCM) outputs from Hadley centre model, the HadRM3 were used. These outputs are for the future term of 2021-2050, for the IPCC A1B scenario. This is also labeled as "2035", for brevity (which is the median of the period).
- IPCC A1B scenario. The A1 storyline describes the future as a world

of very rapid economic growth, a declining global population after it peaks in mid century and the rapid introduction of new and efficient technologies. Major underlying themes are convergence among regions, capacity building and increased cultural and social interactions, with a substantial reduction in regional differences in per capita income. The A1 scenario family develops into three groups that describe alternative directions of technological change in the energy system. The three A1 groups are distinguished by their technological emphasis: fossil intensive (A1FI), non-fossil energy sources (A1T), or a balance across all sources (A1B) (where balanced is defined as not relying too heavily on one particular energy source, on the assumption that similar improvement rates apply to all energy supply and end-use technologies).

IMPACT OF CLIMATE CHANGE ON THE WATER SECTOR

Krishna Basin

Figures 5.3a and 5.3b show the percentage change in precipitation by 2021-2050, with respect to the baseline for the Kharif and Rabi seasons. For the Kharif months, it can be observed that for districts on the southern part of Krishna basin, the rainfall is projected to increase by as much as 10 per cent. For districts like Shimoga, Chikmaglur, Haveri and Uttara Kannada and Dharwad, the rainfall is expected to increase between 20 per cent and 10 per cent.

For Rabi season, a decline in rainfall is projected for most districts in the Krishna basin. The Yadgir district, in the North-Eastern part of Krishna basin is projected to show a decrease in rainfall by 20 per cent, whereas in districts like Gulbarga, Bellary and Raichur, the percentage increase in rainfall ranges from 0 to 10 per cent.

Cauvery Basin

The Figures 5.3c and 5.3d show the percentage change in precipitation for the Kharif or monsoon and Rabi or post monsoon seasons, respectively for the Cauvery basin. During the Kharif season, percentage decline in rainfall is projected for all districts in the Cauvery basin with upto 20 per cent decline projected for Kodagu and Chikmaglur districts. In the remaining districts of the Cauvery basin, the decline in rainfall ranges from no change to 10 per cent. This is likely to have serious implications for Kharif crops.

For the Rabi or post monsoon season, a decline in rainfall by over 20 per cent has been projected for Chamrajnagar. In the case of Ramnagar and Bangalore Urban districts, a rainfall decline of 20 to 10 per cent has been projected. In Mandya and Mysore districts, no change in rainfall to 10 per cent decrease has been projected. Rainfall in Hassan, Kodagu, Tumkur and Chikmaglur districts is projected to increase by 10 per cent in the future.

EVAPOTRANSPIRATION

Krishna Basin

The percentage increase in evapotranspiration for the Kharif and Rabi seasons. The color scale ranges from red (between 20 and 25 per cent decrease in runoff) and orange (between 15 per cent to10 per cent decrease in runoff) to light brown (10 to 15 per cent decrease in runoff), beige (upto 5 per cent decrease in runoff) and light green (between 0 per cent and 5 per cent increase in runoff). It can be observed that in both seasons, in most of the districts, there is a decrease in evapotranspiration in the Krishna basin during Kharif as well as Rabi seasons.

In Kharif season, the South-Western districts are projected to show an increase in evapotranspiration. In the districts of Shimoga, Chikmaglur, Hassan and Haveri, ET is projected to increase in the range of 5-10 per cent by 2021-2050 in the Kharif season. Conversely, for the same districts, the ET is projected to reduce in the Rabi season. The evapotranspiration of Raichur district on the east of Krishna basin is expected to decline by 20 per cent in Kharif season and by more than 20 per cent for Rabi season.

Cauvery Basin

The percentage change in evapotranspiration under the A1B scenario for Cauvery basin. Overall in the Kharif season, evapotranspiration is projected to decrease. Chamarajanagar and Mandya districts are projected to show a decline in ET, whereas Kodagu, Hassan and Chikmaglur are projected to show an increase in ET. The remaining districts are projected to show marginal decrease of 0-5 per cent. The orange color indicates a significant decline in ET while the dark green color indicates an increase in ET. For the Rabi season, only Mysore is projected to show a significant decline in evapotranspiration, and Ramanagar is projected to show maximum increase in ET. Mandya and Tumkur also are projected to show increase, while the remaining districts are projected to show marginal decrease in ET. The red color indicates a significant decline in ET while the light green color indicates an increase in ET.

RUNOFF

Krishna basin

The districts in the Krishna basin and the impact of climate change on total runoff during the monsoon/Kharif season. The color scale ranges from red (upto 25 per cent decrease in runoff) and pink (between 25 per cent-0 per cent decrease in runoff) to light green (between 0 per cent and 25 per cent increase in runoff) and dark green (between 25 per cent and 50 per cent increase in runoff). It can be observed that during Kharif season, an increase in runoff by more than 50 per cent is projected for the northern most district of Bidar. For districts on the south of Krishna basin increase in runoff ranging between 0 and 25 per cent is projected. It is observed that for the remaining districts in Krishna basin, runoff is not expected to increase significantly, especially for Bagalkote and Bijapur districts where a decline in runoff is predicted. The overall increase in runoff is likely to have serious implications for rainfed crops in these districts. The increase in total runoff for Krishna basin post monsoon/Rabi season. It can be observed that a decline in runoff is predicted for the northern and central districts with the exception of Bidar where, along with the districts on the south of the Krishna basin an increase in runoff is projected.

Cauvery Basin

The percentage change in total runoff for Cauvery basin. Red colour represents significant decrease and green colour represents increase. For both the Kharif and Rabi season, there is an increase in runoff in the Eastern districts of the basin. For Kharif season, districts like Mandya, Chamaraja Nagar and Ramanagar is projected to show increase in runoff by 50 per cent, whereas districts like Hassan and Mysore represents a decline by 0-25 per cent.

For the Rabi season, Chamaraja Nagar, Mandya, Ramanagar, Bangalore Urban and Bangalore Rural has been predicted to show an increase in total runoff in the A1B scenario. Tumkur and Chickmagalur have been predicted to show no change to an increase in 25 per cent runoff. However Hassan, Mysore and Kodagu are projected to show a decrease in the runoff, upto 25 per cent, when compared to baseline scenario. The increase in runoff in eastern districts in the Cauvery basin is a cause of concern.

WATER YIELD

Krishna Basin

Water yield is the amount of water leaving a watershed during a specified time period. It can be used as a proxy variable for measuring runoff. The data for water yield was obtained from IIT Delhi, and was prepared using SWAT model. For Krishna basin, a marginal increase in the average water yield has been projected in the mid-century scenario, when compared to the baseline scenario.

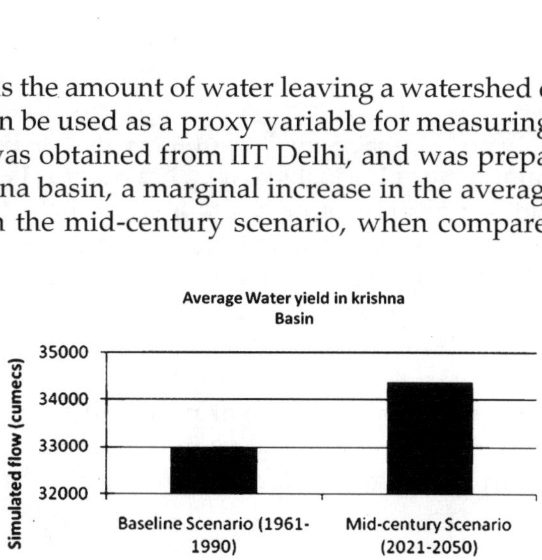

Fig. Water yield in Krishna basin

Cauvery Basin

For projections of the climate change scenario (2021-50's), the average water yield decreases in the Cauvery basin as compared to the baseline. This indicates high amount of water not retained, in the basin. This is of immense concern to farmers, as this water scarcity will manifest itself as reduced yields in the Cauvery basin.

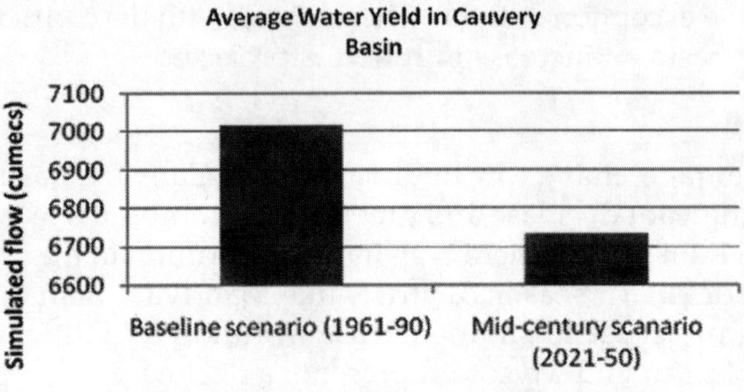

Fig. Water yield in Cauvery basin

DROUGHT

Drought is a normal, recurrent feature of climate and is observed in all the climatic zones. However, it has significantly different characteristics from one region to another. Drought over a geographic area is a temporary condition caused by significantly less (deficient) rainfall for an extended period of time, usually during a season when substantial rainfall is expected over the area. The severity of the drought can be aggravated by other climatic factors such as high temperature, high wind and low humidity.

Krishna Basin

The drought prone districts in Krishna basin are Belgaum, Bellary, Bijapur, Chikmagalur, Chitradurga, Dharwar, Gulbarga, Hassan, Raichur and Tumkur (Compendium of Environment Statistics, 2002). It is projected that during the monsoon season, Raichur will be most affected with 10 to 20 per cent decrease in rainfall projected for the future while the other drought prone districts show a marginal decrease in rainfall (0-10 per cent). The Chikmagalur and Dharwar districts show an increase in rainfall during the same period. During the post-monsoon season, Raichur, Gulbarga and Bellary show a marginal decrease where as other drought prone areas show an increase in rainfall. The decrease in rainfall in these regions, especially in monsoon, will increase the water stress which may affect the crop yields negatively.

Cauvery Basin

The drought prone districts in Cauvery basin are Bangalore, Chikmagalur,

Hassan, Mandya, Mysore and Tumkur (Compendium of Environment Statistics, 2002). During the monsoon season (Kharif) all the districts excepting Chikmagalur show a marginal decrease (0-10 per cent) in rainfall, only Chikmagalur shows a marginal increase in rainfall (0-10 per cent). During the post-monsoon season Bangalore, Mandya and Mysore show a decrease in rainfall, where as Chikmagalur, Hassan and Tumkur show a marginal increase in rainfall during the same period. The decline in precipitation during monsoon can affect the crop growth due to increased water stress.

An investigation of the percentage changes of the three variables - ET, Runoff, and Precipitation under the A1B climate change scenario, when compared to the baseline is conducted for both Krishna and Cauvery basins.

Krishna Basin

Precipitation: Figure indicates that on an average, annually, there would be a decrease in precipitation in the A1B scenario. Investigating further, there is a decrease in precipitation in the monsoon/Kharif season in the A1B scenario, whereas there is an increase in precipitation in the post monsoon/Rabi season, however, percentage-wise, the increase and decrease is very low.

RUNOFF

There is an increase in annual runoff in the Krishna basin, the percentage increase in runoff annually, is on an average 3 per cent, in the A1B scenario. There is high increase in runoff in the monsoon/Kharif season (around 5 per cent, average). However it can be observed that there is a decrease in runoff in the post-monsoon/Rabi season (around 2 per cent, average).

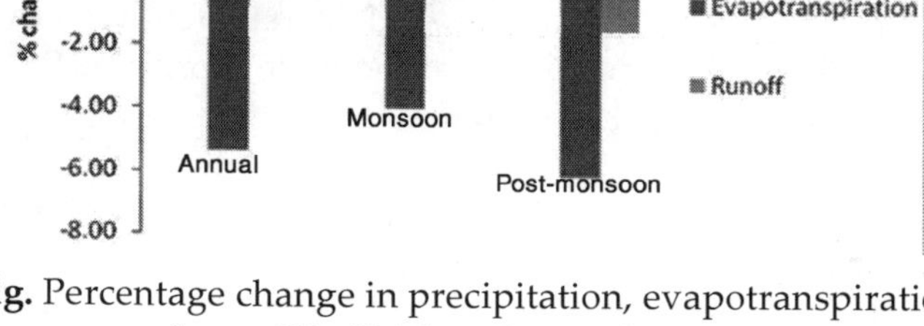

Fig. Percentage change in precipitation, evapotranspiration and runoff in Krishna Basin of Karnataka

EVAPOTRANSPIRATION

There is a decrease in the rate of evapotranspiration in the Krishna basin. Specifically an annual decrease of 6 per cent is observed, the Kharif season decrease is on an average around 4 per cent, while the Rabi season decrease is higher in absolute value, and around 6 per cent on an average. This evapotranspiration decrease may be indicative of water stress in the climate change A1B scenario.

Cauvery basin

Precipitation: On an average, a slight decrease in precipitation annually in the Cauvery basin is projected under the A1B scenario. There is a decrease in precipitation in the monsoon season, an average of around 0.25 per cent in the A1B scenario, when compared to the baseline scenario. However there is an increase in precipitation in the Rabi season in the A1B scenario. Even though the decrease in precipitation is slight, the decrease in evapotranspiration and the increase in runoff is a cause of concern, contributing to the uncertainty about the yields in the future in the Cauvery basin.

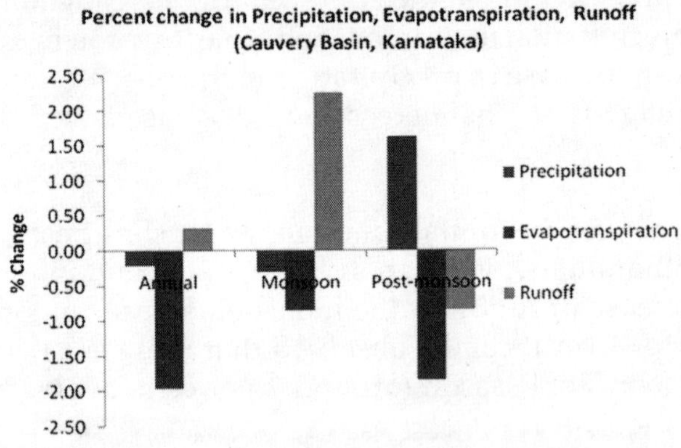

Fig. Percentage change in precipitation, evapotranspiration and runoff in Cauvery basin of Karnataka

Runoff: There is an increase in run off annually in the A1B scenario, when compared to the baseline scenario. The average annual increase in runoff is around 0.25 per cent. The runoff increases during the monsoon season, an average of around 2.2.5 per cent in the A1B scenario, when compared to the baseline scenario. The runoff decreases in the post-monsoon/Rabi season (0.75 per cent in an average) when compared to the baseline scenario.

Evapotranspiration: It can be observed that there is a decrease (2 per cent) in evapotranspiration annually, in the A1B scenario, when compared to the baseline scenario. The decrease in evapotranspiration on an average is around 0.75 per cent in the monsoon/ Kharif annually, when compared to the baseline scenario. The decrease in evapotranspiration was around 2 per cent in the post-monsoon/Rabi season. This decrease in evapotranspiration in all the seasons is indicative of water stress and is a cause of concern in the Cauvery watershed. Overall, among the two river basins, the increase in runoff and decrease in evapotranspiration in the Krishna basin in the A1B scenario is higher in absolute terms when compared to the Cauvery basin on an average, implying higher water stress and potential of droughts in the Krishna basin in the A1B scenario.

EFFECT OF CLIMATE CHANGE ON WATER RESOURCES OF KARNATAKA AND IMPACTS ON CROP YIELDS

Krishna basin

Due to basin closure and high irrigation development in the Krishna basin, the streamflow to the oceans is only 20 per cent of pre-irrigation discharge of the river. There are competing demands for irrigation from the tributaries, which is causing increasing basin closure in the tributaries as well. The increasing source of irrigation for crops in this location is groundwater. The major results of this study predict that there will be a decrease in precipitation, slight increase in runoff and water yield and decrease in evapotranspiration in the short–term (2020-2050). This is indicative of additional water stress in the future. The major crops cultivated in Krishna basin are rice and other grains, sugarcane, banana and other cash crops. These are all water-intensive in nature of cultivation, and therefore the increasing water stress is predicted to have negative effect on these crops, creating possible changes in cropping pattern and species cultivated.

Cauvery basin

The results of the assessment of the impact of climate change on the Cauvery basin indicates that decrease in precipitation in the basin, slight increase in runoff and water yield and a decrease in the evapotranspiration. This will affect the crops that are cultivated in the region. The major crops grown in the Cauvery region are rice, sugarcane, ragi, jowar and cash crops such as coffee, pepper and banana. The staple crops such as rice, and the other crops such as sugarcane and the cash crops are water-intensive in cultivation. Thus the increasing water stress in the region is predicted to affect the crops in the region negatively, leading to changes in the species cultivated.

8

Drought Resistance

INTRODUCTION

Plant responses to drought stress are very complex as stress itself involves various climatic, soil and agronomic factors, frequently complicated by substantial variation in timing of occurrence, duration and intensity. The general complexity of drought is often aggravated under rainfed conditions in marginal areas by erratic and unpredictable rainfall, and by the occurrence of high temperatures, high levels of solar radiation, and poor soil characteristics. This high variability in the nature of drought stress, aggravated by an insufficient understanding of its complexity, has made it difficult to identify specific physiological traits required for improved crop performance under drought, consequently limiting plant breeding efforts to enhance crop drought tolerance. It is highly probable that optimal drought-adaptation requires the combination of several morphological, physiological, and phenological processes, which depend on a multitude of genes and varies within each target environment. In addition to its direct effects on yield, drought can also reduce the potential beneficial effects of improved crop management practices such as fertilizer application or pest and disease management. Given the increasing scarcity of water resources, and competition for them, irrigation is not a practical option for alleviating drought in most of the rainfed areas. Drought management strategies therefore need to focus on maximizing extraction of available soil moisture and the efficiency of its use in crop establishment, growth, biomass and grain yield.

However, agronomic and genetic options that do not involve external inputs of irrigation can only partially alleviate drought effects, because yield will always be lower than what can be achieved with irrigation. This paper will discuss, from a physiological and breeding perspective, some of the current challenges and opportunities in genetic enhancement of drought resistance in rice and introduce some of the screening tools and methods.

Challenges in manipulating drought resistance (DR) genes:

- Most physiological and metabolic processes are affected by water deficits: cell growth, stomata, photosynthesis, translocation, etc..

- A very large number of genes regulated by drought
- Large genetic populations required
- Even component traits are difficult to screen
- Logistical challenges for dealing with a large number of genes even when identified

WHAT IS DROUGHT RESISTANCE?

It is the ability of a plant to maintain favorable water balance and turgidity even exposed to drought conditions there by avoiding stress and its consequences. Stress avoidance due to morphological anatomical characteristics which themselves are the consequences of the physiological processes induced by drought these zerophytic characteristics are quantitative and vary according to environmental conditions. A favorable water balance under drought conditions can be achieved by transpiration before as soon as stress is experienced. These are called "water savers" or. Accelerating water uptake sufficiently so as to replenish the lost water called as "water spenders"

The Mechanism for Conserving Water:-

- Stomatal mechanism: -Stomata of different species vary widely in their normal behaviour and range. In some species stomata remain open continuously or remain closed continuously. Many cereals open their stomata only during a short time in the early morning and remain closed during rest of the day. There is a difference in this respect between varieties of the same crop as shown by the example in two varieties of oat one is more resistant to drought open its stomata more rapidly in the early morning when moisture stress is at its minimum and photosynthesis can precede with the least loss of water.

  However mechanism of conserving water based on the closure of stomata will inevitable load to reduce photosynthesis and may lead to drought induced starvation injury.
- Increased/ Photosynthetic efficiency:- On possibility for overcoming limitations on photosynthesis, imposed bicoastal closure as means for increasing resistance to loss of water by transpiration there by transpiration there by accumulations of CO_2 would be at higher rate for a given stomatal opening. A number of imperfect crop plants (maize, sugarcane sorghum prose, fox tail and finger millets) as well as certain forage species Bermuda grass (Cynodon dactyl on) Sudan grass Bahia grass (Paspalum notatum) Rhodes grass (chloris Guyana) and certain A triplex sp. fixed most of CO_2 into the C4 of molic and aspartic acids so called C4 dicarboxylic acid (C4) pathway.
- Low rate of cuticular transpiration: - The typical example is the cacturs. Thick cuticle results in low rate of transpiration.

- Decreasing transpiration by a deposit of lipids layers on the surface of the leaves on exposure to moderate drought e.g. soybean.
- Reduce leaf area: - The principal means of reducing water loss of xenomorphic plants is their ability to reduce their transpiring surface. Apart from the common means of keeping the aerial parts small perhaps the simplest form of this reduction of the transpiring surface is the sealing or of leaves at the time of water stress a characteristic phenomenon exhibited by many grasses. The rolling of leaves has been shown to reduce transpiration by almost 55 percent in semi conditions and by 75 percent in desert xerophytes.
- Leaf surface: - Various morphological characteristics of leaves he reduce the transpiration rate and may affect survival of plants drought conditions. Leaves with thick cuticle waxy surface and the presence of spines etc. are common and effective.
- Stomatal frequency and location: - A smaller number of stomata retard the development of water deficits. In certain species, the stom are located in depression or cavity in the leaves which is feature can further reduce transpiration by limiting the impingement of currents.
- Effect of awns: - Awned varieties of wheat predominate in the drier at warmer regions and have been found to yield better than awnless one especially under drought conditions though there are exceptions. Awns have chloroplasts stomata and so as photosynthesized. It has been found that the contribution of the away to the total dry weight matter of the kernels was 12 per cent of that the entire plant.

To Improving Water Uptake

- *Efficient root system:* The root systems of drought resistant plants are characterized by wide variety of apparent adaptations. These responded to such predominant soil conditions as the duration of soil dryness and the depth that is normally wet. Plants become adapted to dry conditions mainly by developing an extensive root system rather that structural modification of the roots (shields - 1958). The conceres "extensive root system" includes additional growth of secondary hair roots.
- *High root to top ratio (R/T):* A high root to top ratio is very effective mean to adoption of plants to dry conditions of the growth rate of the roots considerably exceeds that of the shoots. The transpiring surface is there by reduced while root system of the individual plant obtains it's water from a large volume of soil has shown that an increased root top ratio may actually result in greater amount of total dry matter of plants grown under dry conditions as compared a similar ones grown with full moisture.

- *Difference in osmotic potential of plants:* Levitt (1958) has calculated a difference of 0.5 per cent in soil moisture content that includes per manual wilting could supply a plant with enough water to keep it alive for 6 days. This could mean in certain cases the difference between survival and death.
- *Conservation of water spenders to water stress:* Because of increased water absorption water spenders are characterized by very high rate transpiration. However as soon as the absorption rate becomes insufficient to keep up with water loss the water spenders generally develop some of the characteristics of the water savers.

Mitigating Stress

- *Mitigating stress:* Adoptions a drought basis mitigating effects of stress permit the plant to maintain a high internal water potential inspite of drought conditions. They therefore able to maintain cell tartar and growth avoid direct or indirect metabolic injury due to dehydration.

Drought Tolerance

When plant is actually submitted to low water potential it can show drought tolerance by either mitigating the actual stress induced by the moisture deficiencies or by showing high degree of tolerance to stresses.

- *High degree tolerance; Resistance to dehydration:* The simplest method of avoiding drought induced damage is by resisting dehydration, preferably tot he extent.of maintaining turgur and at least by avoiding cell collapse after loss of turgur retain their turgur and therefore can continue to grow when exposed to drought stress. When plants are grown in their natural environment their osmotic potentials tend to be characteristic for each ecological group.

CHARACTERIZATION OF THE TARGET ENVIRONMENT AND DROUGHT DEFINITION

Importance of timing, duration and intensity of stress occurrence. Use historical climate series, crop simulation and water balance models to define the breeding target: Germination, crop establishment, vegetative growth and tillering, flowering, grain filling.

The functional definition of 'drought resistance' should be based on yield stability under water defictits. Drought resistance/tolerance has to be researchable, breakable into simpler levels (e.g. floret sterility and ASI in maize, rooting effects vs. rooting pattern..), and should focus on heritable genetic variation, linked to yield, feasible

Is rice more drought sensitive than other crops/cereals? Rice has generally a Lowland and semi-aquatic adaptation, a shallow compact root system (especially in lowlands), Deep roots not very efficient at water extraction,

Higher tissue sensitivity to water deficit? High sensitivity of grain set to stress, Very sensitive to timing of stress, Other mechanisms?

Drought screening environments

- Artificially create stress in the "normal" growing season: avoids genotype x season problems, excluding water is costly/difficult (reproducibility). But no control of stress occurrence (timing, severity and duration)
- Manage water-deficits in a non-growing season: application of uniform, repeatable and controlled stress environment, possibility to maximize genetic component of the observed variation. BUT the extrapolation of results to 'natural' target environment can be difficult.

The main objectives of a screening system should be:

- Focus on target environment and adaptation to major stress constraints
- Minimize problems in detecting heritable differences in drought resistance (DR)
- Yield under stress is as function of: yield potential, escape, and drought response, therefore the use drought sensitivity index (DSI) can help to distinguish DR from escape and yield potential
- Ultimately, DR should be viewed as the ability to maintain yield components in stress compared to non-stress: maintenance of grain number, grain size, biomass, yield, harvest index.

In addition, screening system should: (i) differentiate among genotypes: degree of yield reduction (+ 50 per cent), manage 'escape' effects, use period of max evapotranspiration (ET) demand, time initiation of stress carefully, adjust for differences in escape, (ii) apply uniform stress (drought nursery approach): "uniform" soil water profile, uniform pre-stress crop growth, water application and water use, (iii) be repeatable: with standard planting time and management procedures, initiate stress by phenological stage, stable water use environment, dedicated field facility, well-established management practices

Screening systems should also take into account the following parameters/ criteria: variations in crop phenology and morphology, the growth stage at which stress occurs, the severity and duration of stress, the management options and levels (e.g. fertilizer applications), the plot size and experimental design, all potential interacting factors such as biotic and other abiotic stresses (e.g. temperature), crop quality and economic factors Control of field irrigation:

Method	Freq	Advantages	Disadvantages
Surface/ Furrow	variable	Robust components	Inefficient water use
	3 x wk	Efficient water use	Requires land leveling

		Easy to measure	Hard to measure amount of water Requires raised beds Row spacing too wide for rice
Sprinkler	3 x wk	Robust components Efficient water use Easy to measure	Uneven distribution pattern same timing of stress for all entries Large border areas needed Leaks, blockages and wind
Drip	daily	Differential irrigation/ plot Efficient water use Uniform application	Expense of components (filters etc) Limits weed control options Labor cost: check quality of irrigation

Screening Tools and Methodologies

- Line-source irrigation
- Rain-out shelters
- Drained paddy (lowland)
- Continuous intermittent stress (upland)

In order to identify sources of drought resistance it is necessary to develop screening methods that are simple and reproducible under the target environment conditions. Therefore, managing drought-screening nurseries requires a careful analysis of likely sources of non-genetic variation among plots, replications and repeated experiments, and establishing procedures for minimizing these factors. Several field and laboratory screening methods have been used successfully to screen for drought resistance, including line-source sprinkler irrigation, rainout shelters, and measurement of drought susceptibility index (DSI). The line-source sprinkler irrigation method creates a gradient of drought stress, and allows the evaluation of large numbers of genotypes at varying intensities of drought in a given environment. However, where response to applied water is linear, simpler stress/no stress techniques provide a more efficient, means of conducting preliminary evaluations. The drought susceptibility index (DSI) proposed by Fisher and Maurer (1978), for cases where the yield performance under stress was not related to days to 50 per cent flowering, was calculated based on yield under rainfed conditions and potential yield under irrigated conditions. The lower the DSI, the greater is the drought tolerance of the line.

DROUGHT RESISTANCE IN THE HOME LANDSCAPE

In most landscapes there are plants that survive periods of drought. Placing these plants in the garden reduces the need to supply extra water during periods of inadequate rainfall. While cacti and succulents may have a place in some drought-tolerant gardens, they are not the only alternatives. There are other choices available and appropriate for dry areas. With careful selection, planning and execution, drought-tolerant landscapes can be as pleasing as those needing heavy irrigation.

SELECT PLANTS FOR THE GROWING CONDITIONS

Select plants for the growing conditions in a given area when planning and designing the landscape. These growing conditions create microclimates. Define the microclimates. Create a site plan that groups plants according to similar water needs. Strive for three basic divisions: very low water zone, low water zone, and moderate water zone. Each area should be irrigated separately, according to specific water needs. Incorporate leaf size and color, bloom period, and size and shape of plant, to create visual interest.

Soil type, wind, and exposure to sun can create a dry microclimate in an area that receives heavy rainfall. Sandy, well-drained soil will dry out more quickly, while heavy clay soils are likely to remain moist much longer. Mixing humus into the soil will improve water-holding capacity of sandy soils and water penetration in clay soils. Sandy soils with humus will hold larger amounts of water. Clay soils amended with humus will absorb more water, reducing runoff. In locations with poor drainage and heavy soils, many drought-tolerant species may suffer during years with average or abundant rainfall. Wet winters and spring rains may also cause problems for some drought-tolerant plants growing in heavy soils.

Limit Turfgrass

Turfgrass requires more water than other plants in the landscape. By isolating turf from gardens, trees and shrubs, a gardener can separate irrigation zones and waste less water. Limit turf to areas that are regular in shape and easily irrigated. Cultivar selection should be appropriate to the climate, site, level of maintenance, intended use and reduced water consumption. Although coarse in texture and appearance, tall fescue mixes offer good heat and drought tolerance. Cut the grass at a height of 21/2 to 3 inches. Longer leaf blades help to shade and cool the ground, reducing evaporation from the soil and lessening the need to irrigate.

Irrigate Efficiently

Efficient irrigation may mean including an irrigation system. The least efficient system is the sprinkler. It delivers a large amount of water in a short period, but loses excessive amounts of moisture to evaporation. Sprinklers are the only choice for turf areas.

Low-volume trickle or drip irrigators and soaker hoses deliver moisture over a long period, losing little water due to evaporation or runoff. Check all systems regularly. An improperly calibrated, clogged or leaking system can waste a great deal of water. Carefully probe the root zone to help determine the moisture content in the soil. With some low-volume systems, the surface of the soil will not appear to be saturated, while the root zone will receive the proper amount of moisture.

In times of drought, utilizing recycled household or graywater helps ease water usage. Graywater should be free from oils, foodscraps and bleach. Mixing graywater and fresh water (half and half) is ideal. Graywater should not be used on root crops or potted plants or be a plant's sole source of water. In addition to graywater, saving rainwater from roofs in cisterns and rain barrels will lessen the need to use community or well water. The need for irrigating an area will depend on all the factors of culture and microclimate. Well established gardens will require less supplemental irrigation during drought than newly planted areas.

Use Mulch Properly

Properly mulching an area lowers the soil temperature and decreases the loss of moisture due to evaporation. In addition to creating texture in the landscape, organic mulches decay, adding nutrients to the soil. Appropriate depth of the mulch is important: 2 to 3 inches for trees and shrubs, and 1 to 2 inches for vegetables, annuals and perennials.

Carefully planned landscapes and sound cultural practices reduce water needs. Controlling weeds will lower moisture competition with other plants. Lessening competition will strengthen existing plants and make them less susceptible to disease, insects and drought. By carefully preparing and meeting plant requirements, a gardener can develop a landscape full of color and texture, while reducing water requirements.

IRRIGATION ZONES

Very Low Water Zone

This area is typically farthest from a source of water. Plants in this area must be chosen carefully, requiring little or no supplemental irrigation. Some of these plants may show problems in years of abundant rainfall.

Low Water Zone

Plants chosen for this area will require more water than that which is available naturally. During severe drought, supplementing the water supply will become necessary.

Moderate Water Zone

This zone will use the greatest ratio of water in the landscape. Keeping this area small will help limit water needs. It is possible to grow drought-intolerant plants in this area.

Plants Selected For Low And Very Low (*) Water Zones

Hardiness may be microclimate dependent. This is not a complete list, but rather a sample of drought-tolerant species. Many species (1) require well-drained soils and will not grow well in poorly drained, clay soils. Some species resistant to drought may perform more vigorously when grown in ideal conditions.

Trees and Shrubs:

- Acanthopanax sieboldiana, Five-Leaved Aralia*
- Acer campestre, Hedge Maple
- A. ginnala, Amur Maple
- Amorpha canescens, Lead Plant
- Aralia elata, Japanese Angelica-Tree
- Arctostaphylos uva-ursi, Bearberry1
- Berberis thunbergii, Japanese Barberry
- Betula populifolia, Gray Birch
- Buddleia alternifolia, Butterfly Bush
- Caragana spp., Peatree
- Carya glabra, Pignut Hickory
- Celtis occidentalis, Hackberry
- Chaenomeles spp., Flowering Quince
- Cotoneaster spp., Cotoneaster
- Crategus phaenopyrum, Washington Hawthorn
- Elaeagnus angustifolia, Russian Olive*
- Genista spp., Woadwaxen
- Gleditsia tricanthos v. inermis, Honey Locust*
- Gymnocladus dioica, Kentucky Coffee Tree
- Juniperus spp., Juniper
- Ligustrum spp., Privet
- Maclura pomifera, Osage Orange*
- Myrica spp., Bayberry*
- Physocarpus spp., Nine-Bark
- Pinus banksiana, Jack Pine
- Potentilla spp., Cinquefoil
- Prunus besseyi, Plum
- Quercus macrocarpa, Burr Oak
- Q. prinus, Chestnut Oak
- Rhus spp. (esp. R. typhina), Sumac*
- Robina pseudoacacia, Locust*

Perennials:

- Rosa rugosa, Hedge Rose*
- R. virginiana, Virginia Rose
- Sophora japonica, Sophora
- Tamarix spp., Tamarix

- Vaccinum pallidum, Dry-Land Blueberry
- Viburnum lentago, Nannyberry
- Achillea spp., Yarrow*
- Aethionema spp., Stonecress1
- Alcea rugosa, Hollyhock
- Alyssum montanum, Alyssum1
- Amsonia jonesii, Amsonia
- Anacyclus depressus, Anacyclus
- Antennaria spp., Pussytoes1
- Anthemis spp., Camomile*1
- Armeria spp., Thrift1
- Artemisia spp., Wormwood, Mugwort*1
- Asclepias tuberosa, Butterfly Milkweed
- Aster divaricatus, White Wood Aster
- A. ericoides, Heath Aster*
- Baptistia australis, False Indigo
- Belamcanda chinensis, Blackberry-lily
- Callirhoe involucrata, Low Poppy Mallow1
- Campanula poscharskyana, Serbian Bellflower1
- Caryopteris x clandonensis, Hybrid Bluebeard
- Centaurea spp., Bachelor's Button
- Cerastium tomentosum, Snow-in-Summer1
- Chrysopsis villosa, Hairy Goldaster*1
- Chrysothamnus spp., Rabbitbush
- Comtonia peregrina, Sweetfern
- Coreopsis verticillata, Tickseed
- Dictamnus albus, Gasplant
- Dracocephalum spp., Dragonhead
- Echinacea purpurea, Coneflower*
- Elymus glauca, Blue Lyme Grass
- Erianthis ravenae, Plume Grass
- Eriogonum umbellatum, Sulfur Plant
- Eryngium spp., Sea-Holly1
- Eupatorium spp., Mistflower, Joe-Pye- Weed
- Euphorbia spp., Spurge1
- Festuca spp., Fescue*
- Gaillardia spp., Blanket Flower
- Guara lindheimeri, Guara1
- Goniolimon tataricum, Tatarian Statice1
- Helianthemum nummularium, Common Sunrose1
- Helictotrichon sempervirens, Blue Oat Grass1
- Hypericum cerastiodes, St.-John's-Wort*
- Hyssopus officinalis, Hyssop

- Knautia spp., Pincushion Flower
- Lavandula spp., Lavender1
- Liatris spp., Gay-Feather1
- Linaria spp., Toadflax
- Malva moschata, Musk Mallow
- Marrubium spp., Horehound
- Nepeta cataria, Catnip
- Oenothera spp., Sundrop1
- Opuntia spp., Prickly Pear*1
- Origanum spp., Oregano1
- Penstemon caeapitosus, Mat Penstemon1
- Perovskia spp., Russian Sage*1
- Potentilla spp., Cinquefoil1
- Ratibida columnifera, Prairie Coneflower*
- Rudbeckia fulgida, cultivars (i. e. Goldstrum)*
- Ruta graveolens, Rue1
- Salvia spp., Sage
- Santolina spp., Lavender-Cotton1
- Scutellaria spp., Scullcap
- Sedum spp., Stonecrop*1
- Sempervivum spp., Hens-and-Chicks*1
- Silene frivaldskyana, Campion
- Solidago spp., Goldenrod
- Stachys spp., Lamb's Ears1
- Symphytum spp., Comfrey*
- Thermopsis fabacea, Carolina Lupine1
- Thymus spp., Thyme1
- Verbascum thapsus, Mullein*1
- Verbena bipinnatifida, Dakota Verbena1
- V. canadensis, Rose Verbena1
- Waldsteinia ternata, Barren-Strawberry
- Yucca spp., Adam's Needle*
- Cosmos bipinnatus, Cosmos
- Eschscholtzia californica, California Poppy*
- Gaillardia pulchella, Blanket Flower
- Gazania rigens, Treasure Flower
- Gomphrena globosa, Globe Amaranth
- Helichrysum bracteatum, Strawflower
- Helipterum roseum, Strawflower
- Ipomoea spp., Morning-Glory
- Portulaca grandiflora, Rock-Rose*
- Sanvitalia procumbens, Creeping Zinnia
- Tagetes spp. and vars., Marigold

- Tithonia rotundifolia, Mexican Sunflower*
- Tropaeolum majus, Nasturtium
- Zinnia spp. and vars., Zinnia

THE DROUGHT BREEDING STRATEGIES

Both conventional and trait-based approaches have been used in breeding programs for drought tolerance. The empirical breeding approach is based on selection for yield and its components in a given drought environment. While such an approach has been partly successful, it requires large investments in land, labor, and capital to screen a large number of progenies plus the difficulty of sampling even a part of the expected range of variability in stress occurrence in the target environment.

In addition, there is evidence of increasingly marginal returns from conventional breeding, suggesting a need to seek more efficient methods for genetic enhancement of drought tolerance. On the other hand, the ability to associate drought adaptive responses with the expression of specific physiological mechanisms has the potential to help greatly in establishing screening protocols and permit better management of genotype × environment (G × E) interactions. However, it has been argued that a focus on very basic mechanisms is likely to be at the cost of the linkages to final grain yield, and increased measurement costs, thereby complicating conventional and molecular breeding for drought tolerance.

Recent research developments in biotechnology have revived interest in targeted drought tolerance breeding and use of new genomics tools to enhance crop drought resistance. Marker-assisted breeding is making possible the improvement of field crops, particularly for traits where phenotyping is only possible late in the season, is difficult, or is prohibitively expensive. As a complement to the recent rapid progress in genomics, a better understanding of physiological mechanisms of drought response will also contribute to the progress of genetic enhancement of crop drought tolerance.

It is now well accepted that the complexity of the drought syndrome can only be tackled with a holistic approach that integrates physiological dissection of crop drought avoidance and tolerance traits using molecular genetic tools such as MAS, microarrays and transgenic crops, with agronomic practices that lead to better conservation and utilization of soil moisture, and better matching of crop genotypes with the environment.

Generally, the response of plants to soil water deficits can be described as a sequence of three successive stages of soil dehydration (Fig. 1). Stage I occurs at high soil moisture when water is still freely available from the soil and both stomatal conductance and water vapor loss are not limited by soil water availability. The transpiration rate during this stage is therefore determined by environmental conditions around the leaves. Stage II starts when the rate of water uptake from the soil cannot match the potential transpiration rate. Stomatal conductance declines, limiting the transpiration rate to a rate similar to that of uptake of soil water, resulting in the maintenance of the water balance of the plant. Finally, stage III begins when the stomata are no longer able to limit the transpiration to that water available from the soil even through stomatal conductance is at a minimum. At this time the plant must resort to other mechanisms of drought adaptation if the plant is to survive.

Crop Yield and Plant Water Use

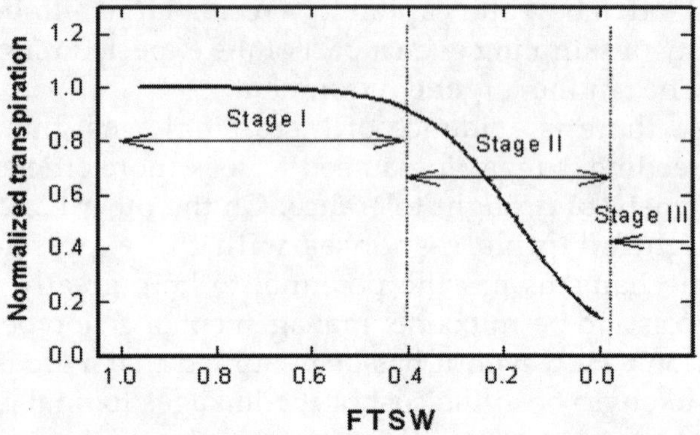

Virtually all major processes contributing to crop yield including leaf photosynthetic rate, leaf expansion and growth are inhibited late in stage I or in stage II of soil drying. At the end of stage II, these growth-supporting processes have reached zero and no further net growth (i.e. increase in biomass) occurs in the plants. The focus of stage III is survival and water conservation mechanisms which will allow the plant to endure under these severe conditions must be used if available. Plant survival is a critical trait in natural dry-land ecosystems, but for most agricultural situations, stage III often, but not always has little relevance to questions about increasing crop yield.

Consequently, the amount of water available up to the end of stage II for all practical purposes determines the cumulative growth and yield on a particular soil. Recovery from stage III can only be of relevance to yield performance if water is added to the system while there is still sufficient time for growth.

Therefore, options involving mechanisms to enhance crop survival, thus do not usually mean any increase in crop yield under severe drought stress conditions. Increased crop yields and water use efficiency generally require the optimization of the physiological processes involved in the critical early stages (mainly stage II) of plant response to soil dehydration.

CHARACTERIZATION OF DROUGHT RESISTANCE TRAITS

Several physiological, morphological and phenological traits/mechanisms have been associated with drought stress adaptation, either in stage-I, stage-II or stage III processes of soil drying including:

- Plant emergence characteristics/vigor
- Phenology/ Elasticity of development
- Nutrient acquisition/Uptake efficiency
- Water use efficiency
- Photosynthesis, Radiation Use Efficiency
- Carbon Discrimination (13C ?)
- Deep Root development
- Hormonal regulation (ABA, GA, Ethylene)
- Osmotic Adjustment/RWC
- Canopy temperature
- Staygreen/ Delayed senescence
- Grain number maintenance
- Grain fill duration and rate
- HI under Drought
- Yield and its components, Etc.

Despite many decades of research on 'drought tolerance' in several crops, little progress has been reported in terms of genetic enhancement of crop productivity under water-deficits environments. Plant physiologists do not always pay enough attention while evaluating the adaptive role of a particular physiological mechanism to grain yield or biomass production under drought. Also, there is no unanimity in relation to what constitutes drought tolerance. Only limited progress has been reported in developing near iso-genic lines (NIL) or recombinant inbred lines (RILS) for specific traits of drought resistance. Another limitation is that lab physiologists have been reluctant to adapt their trait assessment protocols to allow phenotyping of sufficiently large number of individual plants that progenies can be segregated in sufficient numbers for good screening experiments to demonstrate that their target traits are indeed heritable.

Plant breeders on the other hand are unwilling to invest much of their time and resources pursuing traits without this basic information. Breeders and crop physiologists need to work closely in testing the viability/validity of the trait-based approaches for drought tolerance. This has not happened to any great extent previously, thus missing a good opportunity for advancement. Applied breeding programs always want a clear demonstration of the value of a trait before considering incorporating it as a selection criterion. However, without breeder's participation, development of genetic stocks with specific traits incorporated into locally-adapted genetic backgrounds to test the hypothesis of the value of a trait is very difficult.

Identification of simple to observe morphological and phenological traits, reflective of mechanisms and processes that confer drought tolerance is a priority activity in drought research. An appropriate screening trait for drought stress tolerance should fill the following criteria: (i) a strong link with higher or more stable grain yield in the target stress environment, (ii) a high level of heritability, and (iii) the expression of tolerance must be easily measurable, with adequate replication.

INTEGRATED DROUGHT MANAGEMENT OPTIONS

Given the increasing scarcity and competition for water resources, irrigation is generally not a possible option to alleviate drought problems in most rainfed areas. Therefore, drought management strategies, whether agronomic or genetic, therefore need to focus on maximum extraction of available soil moisture and its most efficient use in crop establishment, maximum crop growth, and for increasing biomass and seed yield. In order to plan crop yield improvement programs for a given target drought-prone area the following steps are essential:

- Characterize the major patterns of drought stress and their frequency of occurrence in the target environment.
- Evaluate crop response to the major drought patterns (simulation modeling)
- Match crop phenology (growth period, sowing, flowering, and seed filling) with most favorable period of soil moisture and climatic regimes
- Develop a strategy for the optimal use of supplementary irrigation, when available
- Increase available soil water to crop through agronomic management practices
- Identify plant traits that would maximize (i) use of available soil moisture in transpiration (ii) production of biomass per unit water transpired, and (iii) partitioning into seed, thereby conferring enhanced crop water productivity.

Agronomic and genetic options that do not involve external input of irrigation can only partially alleviate drought effects, because yield is always lower than what can be achieved with irrigation.

9

Remote Sensing for Drought Monitoring

INTRODUCTION

Hydro-meteorological hazards that account for 90 per cent of the natural disasters pose the greatest threat to food safety, the ecological system and economic development by destroying vegetation (including crops and pasture), water resources, livestock and infrastructure. Among these include droughts and floods (trans-boundary phenomena) that have become more frequent in the 21st century in sub-Saharan Africa. The use of environmental satellite data enables us to detect drought and its impacts on agriculture can be detected prior to harvesting using vegetation indices (combination of spectral reflectances) such as NDVI and VCI in Maathuis and KNUST. These indices as proxies of drought and vegetation health/condition are vital in early warning for food security, disaster preparedness and environmental protection. Despite a wealth of remotely sensed data provided by organizations including National Oceanic and Atmospheric administration (NOAA), EUMETSAT, investments in science technology and innovations is often a low priority for decision and policy makers in developing countries such as Eastern Africa region. In addition the countries have limited funds, infrastructure, software and skills to acquire, analyse and interpret the information in real time to near-real time.

OBJECTIVE

The objective of this study is therefore to demonstrate the application of integrated AVHRR and MSG data for drought assessment over eastern Africa as input for monitoring and early warning system for disaster management, at relatively low cost.

Materials and methods

The data used in this study over Eastern Africa were: (a) AVHRR instrument (onboard NOAA series of satellites) 1st dekad September – October – November (SON) season NDVI data, (b) Level 1.5 data of MSG-SEVIRI (Spinning Enhanced Visible and Infrared Imager) for 1st dekad same season

2007, obtained from Institute for Geo-Information Science and Earth Observation (ITC), in Netherlands. The 1st dakad of SON season was selected since it is beginning of rainy/growing season. The MSG data was imported and converted into ILWIS raster format using MSG data retriever with the aid of a Geospatial Data Abstraction Library (GDAL)-driver that reads raw compressed MSG data and facilitates easy geometric and radiometric calibrated data retrieval into formats commonly used by remote sensing packages. Long-term (24 year) maximum NDVI (NDVI max), minimum NDVI (NDVI min) and NDVI for the dekad 1 (first 10 days) of September 2007 (NDVIdek) were computed from the data set. Vegetation condition index for the 1st dekad of September 2007 (VCIdek) was then calculated to determine the vegetation health according to Kogan (1997) in Maathuis and KNUST (October, 2007) hence detect drought which was reclassified according to different severity. VCI expressed in percentage shows how close the current dekad's NDVI is to the long-term minimum NDVI calculated from remotely sensed data.

$$\text{VCI dek} = \frac{(\text{NDVI dek - NDVI min}}{(\text{NDVI max - NDVI min})} * 100\%$$

Kogan criteria	Vegetation condition
>50% - 100%	Optimal
50%	Fair
>35% - 0%	Poor

At 100 per cent VCI, NDVIdek) = NDVI max and at 0 per cent VCI, NDVIdek = NDVI min. According to Kogan (1997), different degrees of drought severity are indicated by VCI below 50 per cent and VCI of 35 per cent is a threshold for extreme drought. Based on this the VCI values obtained were reclassified into drought severity as: 50 per cent to 100 per cent, normal to above normal condition (wet); <50 per cent to 35 per cent, moderate drought; <35 to 20 per cent severe drought; <20 per cent to 0 per cent very severe drought. These scales were arbitrarily selected.

Results and Discussion

The different drought severity according to NDVI derived vegetation condition indices over different parts of Eastern Africa are shown in Figure. It is revealed that some parts of north eastern Ethiopia, south eastern Eritrea and a small area in northern Somalia experienced very severe drought. Most parts of Djibouti and Somalia, south eastern Ethiopia; parts of eastern Kenya, eastern and northern Tanzania and Rwanda had moderate to severe drought. Most parts of Uganda, Burundi, Sudan, western Ethiopia and some parts of western and coastal regions of Kenya had optimal vegetation conditions. The prevailing severe to very severe drought indicates poor vegetation health and likelihood of crop failure in these areas hence having impact on food production. These can also depict the danger of wild fires to human and wildlife/biodiversity as well as environmental health hazards (dust and pollen particles) associated with respiratory diseases. The implications are that if such prevailing conditions are monitored, different stake holders (decision makers, farmers, health officials, wildlife authorities, humanitarian organisations) can be effectively informed about prevailing and/or developing hazardous situation. This is important for food and human security as well as in disaster preparedness, to minimize economic losses for sustainable socio-economic development.

METEOROLOGICAL DROUGHT ANALYSIS USING GIS

Drought is a regional phenomenon, which is considered as one of the major natural environmental hazard, which severely affects the water resources and vegetation through out the history of humankind. Drought indicates the deficit of rainfall, runoff, water availability, soil moisture, ground water, surface water and crop yield that have been the subject of many investigators from the previous century. The National Commission on Agriculture (1976) broadly classified droughts into Meteorological drought, hydrological drought and agricultural drought. Meteorological drought is a situation when the actual rainfall is significantly lower than the climatologically expected rainfall over a wide area. Hydrological drought is associated with marked depletion of surface water and consequent drying up of lakes, rivers, reservoirs etc. Hydrological drought results if meteorological drought is sufficiently prolonged. Agricultural drought is a condition in which there is insufficient soil moisture available to a crop results in reduction of yield. Agricultural drought is the end effect of meteorological and hydrological drought and its severity is assessed from the crop yield. So, the periodical drought assessment in a region must be necessary to take the management measures.

Many researchers related agricultural drought by an inadequate amount of soil water available during the critical crop growth periods. Subramanyam (1967) and Subramanyam et al (1974) considered as a shortage of moisture estimated through simple water budgeting procedure. The index developed for drought assessment was known as Moisture Adequacy Index (MAI), which is ratio of water deficit to over water need. Palmer (1986) defined severity of agricultural drought in terms of the magnitude of computed abnormal evapotranspiration deficit and expressed as a crop moisture index

STUDY AREA

The Krishna basin is situated between East longitudes 730 21' to 810 09' and North latitudes 130 07' to 19025' in the Deccan plateau covering large areas in the States of Maharashtra, Karnataka and Andhra Pradesh. The Krishna basin extends over an area of 2,58,948 sq.km., which is nearly 8 per cent of the total geographical area of the country. Krishna catchment was divided in to 12 sub basins in the report of Krishna Water Dispute Tribunal.

In the present study K-11 sub basin i.e Palleru basin of River Krishna has been considered to analyze meteorological drought at each rain gauge sub area. The Palleru sub basin lies between latitudes 160 39′ and 170 50′ North and longitudes 790 17′ and 800 09′ East. The Palleru sub basin has a catchment area of 3,263km2, which constitutes 1.3 per cent of the total basin area. The catchment area of the Palleru sub basin entirely lies in the State of Andhra Pradesh. In the present study Palleru sub basin has considered up to Palleru reservoir only. Ten rain gauge stations (Palakurthy, Kodagandla, Torrur, Tungaturthy, Aravapally, Maripeda, Noothankal, Atmakur, Kushmanchi and Mothey) were identified in and around the study area. Daily rainfall data was collected for all ten rain gauge stations in the study area for a period of 1992 to 2004 form Bureau of Economics and Statistics, Khairathabad, Hyderabad.

IMD METEOROLOGICAL DROUGHT ASSESSMENT

Thiessen network was developed using Arc GIS 9.1 software. Normal rainfall has been calculated at Palakurthy, Kodagandla, Torrur, Tungaturthy, Aravapally, Maripeda, Noothankal, Atmakur, Kushmanchi and Mothey as 891.73mm, 760.05mm, 865.00mm, 788.60mm, 668.39mm, 576.95mm, 704.45mm, 555.67mm, 869.70mm and 355.52mm respectively. In the sub basin highest rainfall recorded at Noothankal rain gauge station as 676.00mm and lowest is 342.00mm at Kushmanchi rain gauge station. The IMD uses Irrigation Commission report, 1982 to assess the meteorological drought on the basis of percentage deviation of rainfall from the mean rainfall. The percentage deviation is given by the following equation.

$$MD_t = \frac{P_{act} - P_{mean}}{P_{mean}} \times 100$$

where: MDt = Percenatge of meteorological deviation, Pact = Actual rainfall in the time period ‘t’, Pmean = Long term mean rainfall

The percentage deviation of rainfall and the category of drought assessment used by IMD are as given in the Table.1. The IMD drought assessment method was used for ten rain gauge stations in K-11 sub basin of river Krishna. Thirteen water year annual rainfall data average was taken and percentage deviations of each water year and seasonal wise (Northeast and Southwest) percentage deviations were calculated corresponding year and seasonal wise drought severity values for ten rain gauge stations.

Table. Percentage of Deviation and Category of Drought Assessment

Sl.No	Percentage Deviation (MDt)	Class	Category
1	>0	M0	No drought
2	0 to –25	M1	Mild
3	-25 to –50	M2	Moderate
4	<-50	M3	Severe

RESULTS AND DISCUSSION

Based on the results obtained from the IMD drought assessment method for the study area, the drought severity estimates were arrived and discussed below. As seen from the tables, it is observed that K-11 sub basin of river Krishna had experienced mild to moderate drought in the annual assessment for the years 1992, 1993, 1994, 1999, 2000, 2001, 2002, 2003 and 2004. Rainfall in the water years was observed more than the normal rainfall as given in Table.

In the sub basin Mild and moderate drought was observed in the year 1992, 1994, 1999, 2000, 2001, 2002 and 2004 in the entire sub basin. It is observed from the table.3.the water years 1995-96, 1996-97, 1997-98 and 1998-99 was observed no drought. Seasonal assessment Southwest and Northeast monsoon was analyzed for the entire basin for a period of 13 years and given in Table.4 and table5.

Palakurthy sub area was observed no drought in the entire period except 2000 water year in southwest monsoon even in the years where rainfall is less than normal rainfall. But some of the years in the seasons of southwest and northeast monsoons experienced severe, mild and moderate droughts. The spatial variation of drought severity over the basin in a particular year can be depicted in the form of thematic maps. The drought severity values of all the ten rain gauge stations were exported to the GIS software to generate drought severity maps.

IMPACT OF DROUGHT USING REMOTE SENSING

Drought is a complex phenomenon, which is difficult to define. The term is used to refer to deficiency in rainfall, soil moisture, vegetation greenness, ecological conditions or socioeconomic conditions, and different drought types can be inferred. Nevertheless, drought is essentially a climatic phenomenon, a consequence of an abnormal decrease of precipitation. In this study, drought is considered as a period when the precipitation is low in regard to long-term average conditions. Drought periods can result in significant losses to crop yields, increasing the risk of forest fires and triggering processes of land degradation and desertification.

In the present context of climate change and increasing land degradation and desertification, being able to calculate the impact of a drought is crucial in determining the environmental consequences of a hypothetical change in climatic conditions. Indeed, a number of climate models have shown that drought frequency and intensity are likely to increase in several regions.

However, the spatial and temporal complexity of droughts hinders attempts to identify their impact, since drought intensity varies both with the time scale and spatially. The methods used to determine the effects of drought on vegetation are usually based on crop production yields, although tree ring analysis and experimental plots have also been used for this purpose. However, none of these approaches are able to consider continuous information in time and space. Moreover, differences in the effects of a drought between vegetation types cannot be easily analysed with these procedures.

The use of remote sensing data presents a number of advantages when determining drought impact on vegetation. The information covers the whole of a territory and the repetition of images provides multitemporal measurements. In addition, vegetation indexes obtained from satellite data allow areas affected by droughts to be identified (Kogan, 1995 and 1998; McVicar and Jupp, 1998).

In semi-arid lands, several studies provide evidence of the important role of precipitation variability in explaining the vegetation anomalies identified by means of remote sensing. However, the assessment of vegetation vulnerability to water shortages, described by means of drought indexes, has been analysed in only a few studies (i.e., Gutman, 1990; Lotsch et al., 2003). Ji and Peters (2003) have related the evolution in vegetation activity, quantified from remote sensing data, with the drought indexes on the Central Great Plains of the USA, showing a close relationship between both variables. In the same region, Teiszen et al. (1997) have reported a marked reduction in vegetation activity during dry years. Walsh (1987) and Peters et al. (1991) have also analysed the role of drought on vegetation activity by means of remote sensing in other regions of the USA.

Drought indices are particularly useful for monitoring the impact of climate variability on vegetation because the spatial and temporal identification of drought episodes is extremely complex. Moreover, drought impact on vegetation may vary as a function of water use efficiency.

Drought duration may also lead to spatial differences in drought impact on vegetation (Ji and Peters, 2003; Wang et al., 2003), while drought impact can vary as a function of the season of the year, since the water requirements of the vegetation can change markedly over a twelve-month period. Thus, what is required are spatial analyses of drought impact that take into consideration different vegetation types and environmental conditions. Such analyses are enhanced when using remote sensing data.

In the Mediterranean region, some climate models today predict a decrease in precipitation levels during the twenty-first century (Jones et al., 1996, Gibelin and Déqué, 2003), mainly during the winter and summer months. The spatial uncertainty of the predictions is important and the differences between models very high (Houghton et al., 2001), but the models agree that the Mediterranean region would be one of those most affected by the precipitation decrease expected during the twenty-first century.

This will serve to exacerbate drought episodes, land degradation and desertification. Thus, we urgently need to analyse the potential impact of drought on the Mediterranean semiarid ecosystems in order to identify the areas that are most vulnerable to drought. This paper analyses drought impact on vegetation in the northernmost semi-arid region of Europe, the middle Ebro valley. In this area, crop productions and natural vegetation activity are mainly determined by soil water availability. The landscape of the valley is complex, with a variety of land-uses/-covers, while there are major seasonal and spatial differences in the climate. Drought episodes are very intense; with the periods with no precipitations during more than 100 days being common.

The objectives of this study are to determine: (i) spatial differences in the drought impact on vegetation, (ii) whether there are seasonal differences in this impact, and iii) whether the spatial differences in the drought impact are determined by land-cover types, aridity or vegetation characteristics.

METHODS

Remote Sensing data

A range of vegetation indexes based on remote sensing data have been used to monitor vegetation, with the most widely adopted being the Normalized Difference Vegetation Index (NDVI). The data used in compiling the NDVIs are closely related to the radiation absorbed and reflected by vegetation in the photosynthetic processes. There are also close relationships with the vegetation biomass, the percentage of vegetation cover and the leaf area index (Tucker et al. 1981 and 1983; Wylie et al. 2002). Many studies report that the spatial and temporal differences in the NDVI are closely related to the climate in many environments (Eastman and Fulk, 1993; Ichii et al., 2002). In fact, temporal variations in the NDVI may be representative of the vegetation's response to climatic variability (Nicholson et al., 1990; Santos and Negrín, 1997; Potter and Brooks, 1998). Thus, this index has been widely used to monitor ecosystem dynamics and to detect the spatial extent of drought episodes and their impact (Groten and Ocatre, 2002; Tucker and Choudhury, 1987).

In this paper, to analyse drought impact on vegetation, monthly composite NDVI obtained from AVHRR images were used. The NOAA-AVHRR images are received by the Remote Sensing Laboratory of the University of Valladolid (LATUV) at full resolution (1 km2). The images used were obtained from the NOAA-9, -11 and -14 satellites (afternoon passes). The data is received daily and carefully calibrated (Kaufman and Holben 1993; Rao and Chen 1999; NOAA 2003) to avoid artefacts and inhomogeneities in the series caused by the changes of satellites and degradation of orbits. The images are atmospherically corrected by means of a modification of the 5S code considering mid-latitude atmosphere standard parameters. Finally, clouds are eliminated and the NDVI is calculated. Later, the NDVI images are geometrically corrected following Illera et al. (1996). To avoid residual radiometric errors, and to compress the data volume, monthly NDVI composites are created using a maximum NDVI compositing algorithm. Shadows can impact differently on the red and near-infrared bands, and the NDVI could be impacted by this. Nevertheless, although the band ratio (NDVI calculation) does not totally eliminate the problems caused by topography and view-Sun geometry, in low spatial resolution images the problems are not very serious. Burgess et al. (1995) analysed the topographic effects on AVHRR-NDVI considering mountainous spaces. They highlighted that in a complex topographic area errors caused by topography and shadows using 1-km NDVI data are approximately 3 per cent. Therefore, it is reasonable to ignore high-order topographic effects such as sky occlusion and adjacent hill illumination.

Some gaps in the database caused by persistent cloud cover and by the failure of the NOAA-11 satellite between September and December 1994 were filled using highly reliable regression techniques applied to the data from other months. Gaps were lower than 5 per cent of total and the maximum of consecutive pixels/months to be filled were 3 (only 3 per cent of cases to be filled, with the exception of the gap between September and December of 1994). Some months with complete records were selected randomly to test the filling algorithm. The agreement between the real NDVI and the filled data was always higher than R2 = 0.83. A time series of monthly NDVI composites from January 1987 to September 2000 was used for analysis at 1 km2 of spatial resolution.

The NDVI data set developed by LATUV has been tested and widely used in the Iberian Peninsula to monitor vegetation, identify areas affected by drought, to analyse the forest fire risk, to predict crop yields and to determine re-vegetation processes.

The NDVI is determined by factors that include the vegetation type, ecosystem characteristics, phenology, soil, topography. For this reason the NDVI is not comparable in space, especially in non-homogenous areas. Therefore, to analyse the climate impact on vegetation, the NDVI values have to be stratified to avoid the influence of environmental conditions. For this purpose, Kogan (1990) proposed a geographic filter that eliminates the proportion of the spatial variability of the NDVI related to the geographic and ecosystem characteristics. The NDVI value of each pixel is stratified between 0 (minimum NDVI) and 100 (maximum NDVI), yielding a new index denominated Vegetation Condition Index (VCI), which is obtained according to:

$$VCI = 100\frac{(NDVI - NDVI_{MIN,X})}{(NDVI_{MAX,X} - NDVI_{MIN,X})}$$

Where NDVI is the vegetation index, NDVIMIN,X is the minimum NDVI recorded during month X and NDVIMAX is the maximum NDVI during month X. The VCI allows climate impact on different vegetation types and ecosystems to be compared and has been widely used for this purpose in various countries, including the USA, Argentina; China; Australia and Mongolia.

Drought Index Calculation

Various indexes using meteorological data have been used to quantify droughts. However, the most widely used today is the Standardized Precipitation Index (SPI), since it can be calculated at different time scales and hence quantify water deficits of different duration. This index enjoys several advantages over the others. Calculation of the SPI is easier than on more complex indices such as the Palmer Drought Severity Index, because the SPI requires only precipitation data, whereas the PDSI uses several parameters. Moreover, the PDSI has some shortcomings in spatial and temporal comparability. The SPI is comparable in both time and space, and is not affected by geographical or topographical differences.

The SPI has been used for drought analysis in a number of regions (e.g., Hayes et al., 1999; Bonaccorso et al., 2003; Vicente-Serrano et al., 2004c) and for analysing drought impact on crop production yields (Quiring and Papakryiakou 2003; Vicente-Serrano et al., 2006) and remote sensing data (Ji and Peters, 2003; Lotsch et al., 2003).

The SPI is calculated from precipitation data, which means the quality of the index is ultimately linked to the quality of the precipitation series. The longer the series available, the better is the quality of the SPI. Guttman (1999) suggests a minimum of fifty years of monthly precipitation records for calculating an accurate SPI. In this study, we used precipitation data from 41 stations recording data from 1950 to 2000 obtained from the National Institute of Meteorology (Spain). To guarantee the quality of the data set, the series were checked by means of a quality control process that identified anomalous records. The homogeneity of each series was checked by means of the Standard Normal Homogeneity Test. Only one non-homogeneous series was identified, which was corrected following Alexandersson and Moberg (1997). Finally, the remaining temporal gaps were completed using linear regressions with respective reference series.

The SPI is the number of standard deviations that the precipitation summarised during a period deviates from the long-term mean. Since precipitation is not normally distributed, a transformation is applied so that the transformed precipitation values follow a normal distribution. We calculated precipitation at time scales of 3, 6 and 12 months as these scales have been found to be useful for monitoring various drought types (Edwards and McKee, 1997). The SPI was calculated in each observatory according to the Pearson III distribution and the parameters were calculated following the L-moment method (Sankarasubramanian and Srinivasan, 1999). The complete procedure used for the calculation of the SPI is reported in Vicente-Serrano (2006). The drought conditions corresponding to the SPI values, following Agnew (2000), and based on the probability of occurrence of the SPI values. Figure shows the evolution in the SPI at the time scales of 3, 6 and 12 months in Zaragoza (centre of the valley) between 1987 and 2000.

Table. Drought classification according to the SPI values following Agnew (2000)

SPI VALUE	CATEGORY
<1.65	Extremely wet
1.28 to 1.64	Severely wet
0.84 to 1.27	Moderately wet
-0.84 to 0.84	Normal
-1.28 to –0.83	Moderate drought
-1.65 to –1.27	Severe drought
<-1.65	Extreme drought

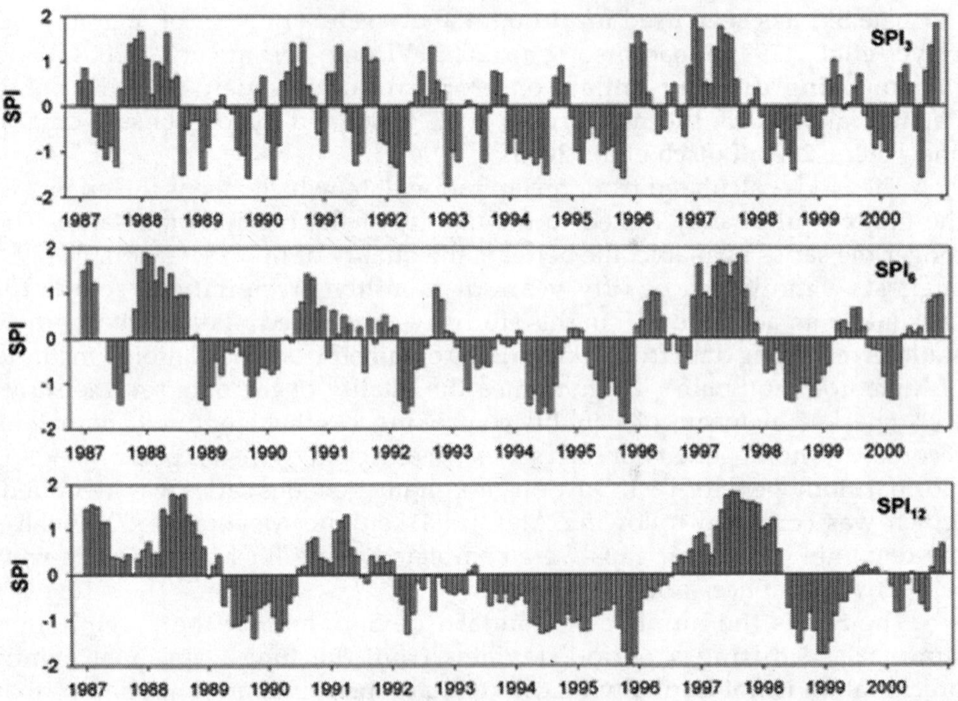

Fig.. Evolution of the SPI in Zaragoza (centre of the valley) between 1987 and 2000

Interpolation of the climatic data

The SPI was calculated at the 41 stations. However, to compare the VCI with the SPI at different time scales throughout the study area, we had to interpolate the SPI values, obtained for each month, at the same spatial resolution as that of the VCI images. To obtain continuous maps of drought indexes, a local method of splines with tension was used, which in the middle Ebro valley has also been found to provide accurate maps of climate averages and time series of maps of drought indexes. The maps were obtained at a spatial resolution of 1 km2 so that they could subsequently be compared with the VCI data set.

Figure shows continuous 3-month SPI maps from January to March of 2000. This particular example shows the spatial complexity of droughts within the study area.

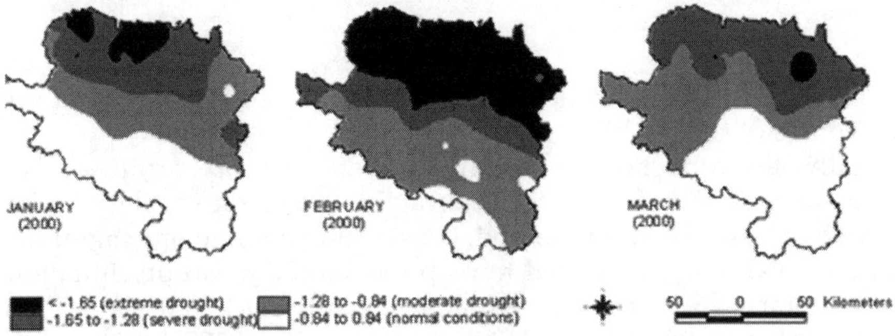

Fig. Continuos maps of 3-months SPI.

To determine the spatial and seasonal variations of the impact of drought on vegetation activity, correlation analysis each 1-km2 pixel was carried out between the monthly SPI series at the time scales of 3, 6 and 12 months (hereinafter SPI3, SPI6 and SPI12) and the VCI series. Thus, the correlation maps indicate spatial differences in the impact of drought on vegetation activity. Correlations were considered significant at $p < 0.05$.

Analysis of the Factors Determining Spatial Differences in Drought Influence on Vegetation

A range of geographic and vegetation variables were used to determine the factors that lead to spatial differences in the influence of drought on vegetation activity. To determine the impact of land cover, the CORINE land cover map (CLC, 1989) was used. The CORINE ('Co-ordination of Information on the Environment) land cover map was an effort of the European Commission (EC) to map the land cover of member states. The CORINE map combines land cover and land use, giving 44 separate classes, in vector, displayed at a scale of 1:100000 with a minimum mappable unit of 25 ha (For further details see e.g., Brown and Fuller, 1996).

The spatial scale of the land cover map was adapted to the spatial scale of the VCI images by means of a mode criterion. Each 1-km2 pixel was assigned to the land-cover type that represented the highest percentage of surface within the pixel. The spatial patterns of the most representative land-cover types within the study area were compared with the spatial distribution of the impact of drought on vegetation 1: Dry farming areas (cereals), 2: Irrigated lands, 3: Deciduous forests, 4: Coniferous forests, 5: Shrubs and pastures.

Additionally, the average NDVI and the Coefficient of Variation (CoV) of NDVI values were used to determine whether there were any differences in drought impact on vegetation as a function of the vegetation activity (described by means of the NDVI) and its temporal variability (described by means of the CoV values):

$$CoV = \frac{\sigma_x}{\overline{x}}$$

Where σ is the standard deviation of the NDVI series in the pixel x and $\overline{x}$ is the average of the NDVI.

Finally, the topoclimatic conditions were also taken into account in understanding spatial differences in drought impact on vegetation. For this purpose, a regression-based method, considering a range of geographic and topographic variables, was used to map the annual mean precipitation and annual air temperature at the same spatial resolution as that of the VCI images (1 km2) (For further details, see Vicente-Serrano et al., 2003).

Aridity is the main climatic factor controlling the spatial distribution of vegetation activity in the study area. Thus, using the regression-based annual precipitation and temperature maps, an Aridity Index (AI) was obtained to summarise the general climatic conditions within the study area and to determine whether the spatial distribution of this factor also led to spatial differences in drought impact on vegetation activity. To obtain a general measure of aridity, an Aridity Index (AI) was calculated based on the ratio between the annual mean precipitation and the annual air temperature maps. This index has been widely used for climatic classification and aridity quantification (Korzun et al., 1976; Botzan et al, 1998). A low index value is indicative of a high degree of aridity.

SEASONALITY OF THE NDVI PATTERNS

The evolution of the average VCI of the study area between 1987 and 2000. The main decrease was recorded in 1991-1992 and 1995-1996 coinciding with the main drought periods that affected this region. On the contrary, between 1987 and 1988 and between 1997 and 2000, although some temporal variability is also shown, a higher vegetation activity is observed, also coinciding with humid conditions.

The spatial and temporal average and standard deviation values of the monthly NDVI for the whole study area. On average, the highest NDVI values were recorded during the spring, with a marked rise occurring in the index between January and April, followed by a slower fall from here to the end of the year. This behaviour is determined by climatic seasonality: in winter, frequent frosts limit vegetation activity whereas in summer, the characteristic aridity reduces vegetation activity.

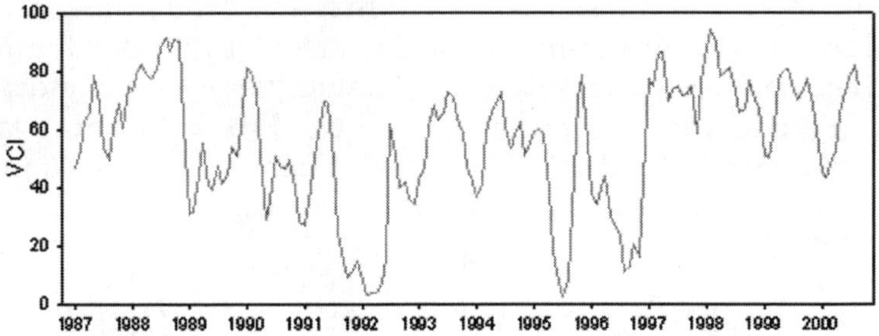

Fig. Evolution of the VCI for the whole study area (1987-2000)

Spring, due to available water and moderate air temperatures, is the season most favourable to vegetation activity and is when the highest NDVI values were recorded. The spatial variability in the NDVI (bars of 1 standard deviation values) was higher during the summer than in winter, because the irrigated lands and the active forests present marked differences in their vegetation activity compared to that of the dry-land agricultural areas and the steppes. By contrast, vegetation activity was very low during winter for all land-cover types and, therefore, a low spatial variability was recorded.

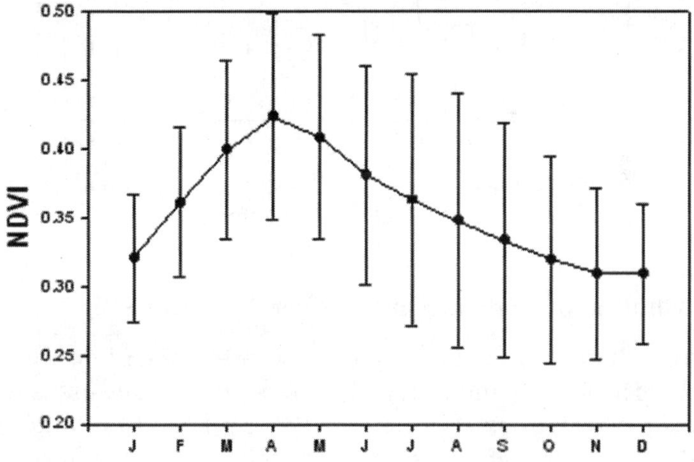

Fig. Mean and standard deviation of monthly NDVI values in the whole study area (1987-2000)

Marked differences in the average monthly NDVI values as a function of land-cover type were recorded. In April, maximum vegetation activity was recorded primarily in the dry-farming areas (cereals). In areas of shrubs, pasture and coniferous forests, seasonal differences were lower. In irrigated lands, the average NDVI values were higher than those recorded in other land-cover types as vegetation activity remained high during the summer months.

The values of Leaf Area Index (LAI) for these land cover types indicate important spatial and seasonal differences. The following values were obtained from MODIS images at a resolution of 1 km2. The dry farming areas have a LAI between 1 and 2 in spring and values close to 0 after the harvest in summer. The coniferous forests have LAI values between 1.5 and 3 in spring and between 2 and 4 in summer. Finally, the shrub and pasture lands have LAI values between 0.5 and 1.5 in spring and between 0 and 0.5 in summer.

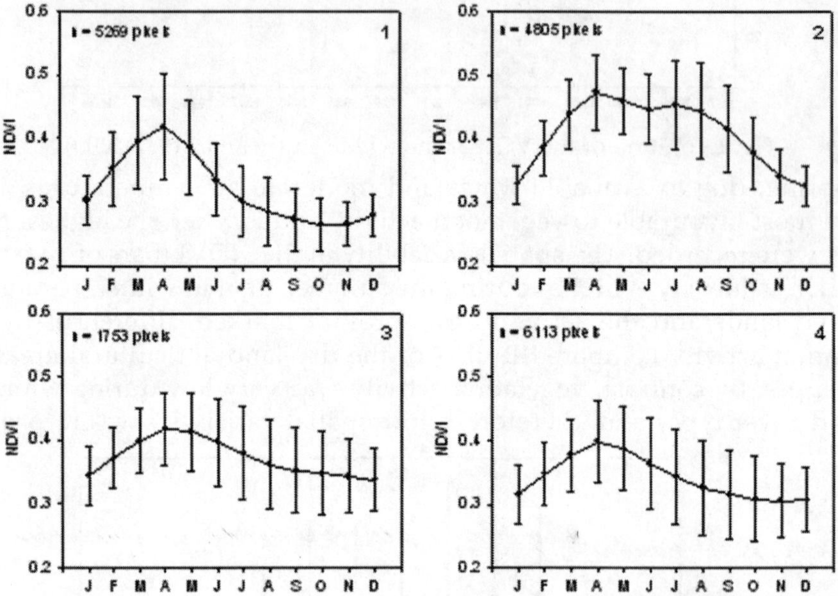

Fig. Monthly averages of NDVI in different land-cover types within the study area (1987-2000). 1. Dry farming areas (cereals), 2. Irrigated lands, 3. Coniferous forests, 4. Shrub and pasture-lands.

Spatial Distribution of Correlations between VCI and SPI

The spatial distribution of correlations between the monthly series of VCI and the series of SPI_3, SPI_6 and SPI_{12}. The black lines isolate areas with positive and significant correlations ($R = 0.53$, $p < 0.05$). For SPI3, large areas to the north of the study area showed significant correlations between the SPI and the VCI in January. These areas are mainly covered by cereals that start to grow in this month. In March and April the correlations were lower. The relationship between SPI3 and the VCI was more significant (lower p values) between May and July, although somewhat patchy, and no consistent spatial patterns according to land-cover type can be identified. The sole exception was the low correlation found in the irrigated lands. Between August and December, the correlations fell dramatically throughout the study area. In this period, the crop-cycle in cereals has finished (July harvesting), while the activity of natural vegetation is reduced because of the fall in temperature.

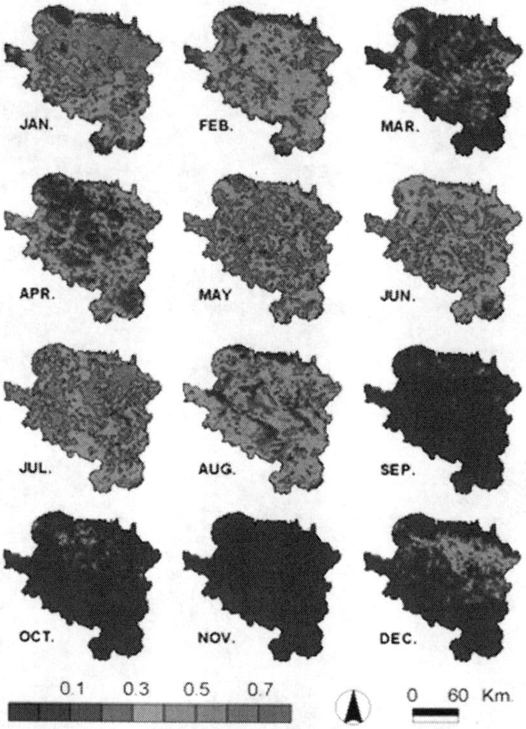

Fig. Spatial distribution of monthly correlations between the VCI and the SPI3. Black lines isolate areas with positive and significant correlations

The monthly correlations between the VCI and SPI6. The seasonality of correlations and spatial patterns change markedly in comparison to those recorded for SPI3. The magnitude of correlations in January and February was not as great. This indicates that in winter, vegetation activity over large areas is more closely determined by the precipitation of the previous 2-3 months than the rainfall recorded over longer periods. By contrast, the percentage surface area presenting correlations with an R-value = 0.53 increased in March and April compared to the figures for SPI3. Between April and June, positive and significant correlations were recorded in the centre of the valley, where the land cover is dry-farming, shrubs and pasture. In July and August, no significant correlations were recorded in the irrigated lands, or in the forests located to the north and south-east. This indicates that in the natural vegetation areas of the centre of the valley, including shrubs, pasture-lands and the coniferous forests, the interannual variability in vegetation activity during the summer is determined mainly by the cumulative precipitation that falls in spring since the correlation between SPI3 and the VCI was very low during these months. In common with SPI3, non-significant correlations between the VCI and SPI6 were recorded during the autumn throughout the study area.

Fig. Spatial distribution of monthly correlations between the VCI and the SPI6. Black lines isolate areas with positive and significant correlations

The spatial distribution of monthly correlations between the VCI and SPI_{12}. Significant correlations were recorded in the spring (April-May) in most of the study area with the exception of the irrigated lands and the forests of the north and south-east. During the spring, the irrigated lands are in the early stages of the vegetation cycles for the majority of cultivations and the needs of water are less important that during the periods of higher evapotranspiration (summer). In the forests located in the mountains of the north and south the higher vegetation activity is recorded later (June-July) and the lower correlations found during the spring months could be related with the differences in phenology in relation to the shrubs, pastures and coniferous forests of the centre of the valley, which are more active during April-May.

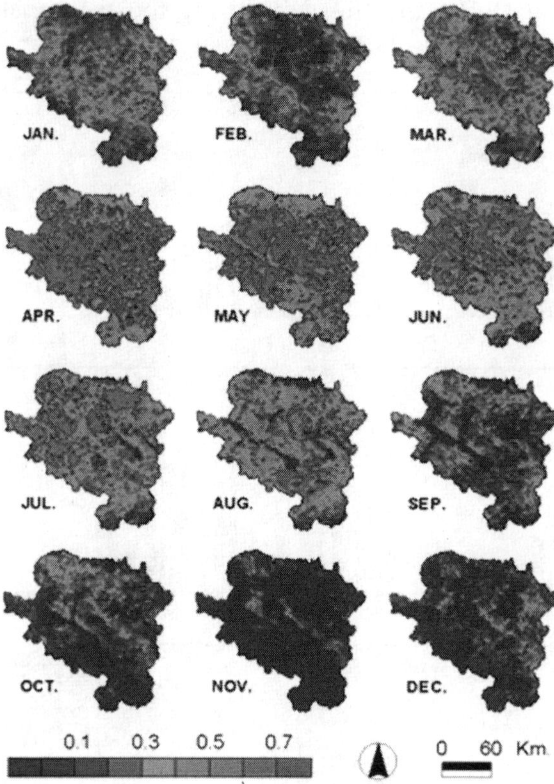

Fig. Spatial distribution of monthly correlations between the VCI and the SPI_{12}. Black lines isolate areas with positive and significant correlations

Spatial Differences in the SPI-VCI Relationships as a Function of the Spatial Distribution of Land-cover types

The percentage of each land-cover type for which significant correlations between the VCI and the SPI were recorded (R-value = 0.53). For SPI3, a large proportion of the dry-farming and irrigated lands presented significant correlations in January and February. Between May and July, about 35-40 per cent of the surface of each land-cover type presented significant correlations, while deciduous forests presented the lowest surface percentage.

For SPI_6, a large proportion of dry-farming areas (>30 per cent) presented significant correlations in April and May between the VCI and the SPI. The percentage surface area with significant correlations increased during the summer for each of the land- cover types. However, marked differences were found as a function of land cover. Dry-farming areas, coniferous forests and shrubs and pastures presented a higher surface area with significant correlations (> 60 per cent of the surface) than did irrigated lands and deciduous forests. For SPI_{12}, large areas presented significant and positive correlations between SPI_{12} and the VCI during the spring, above all areas of dry-farming, coniferous forests and shrubs and pasture-lands.

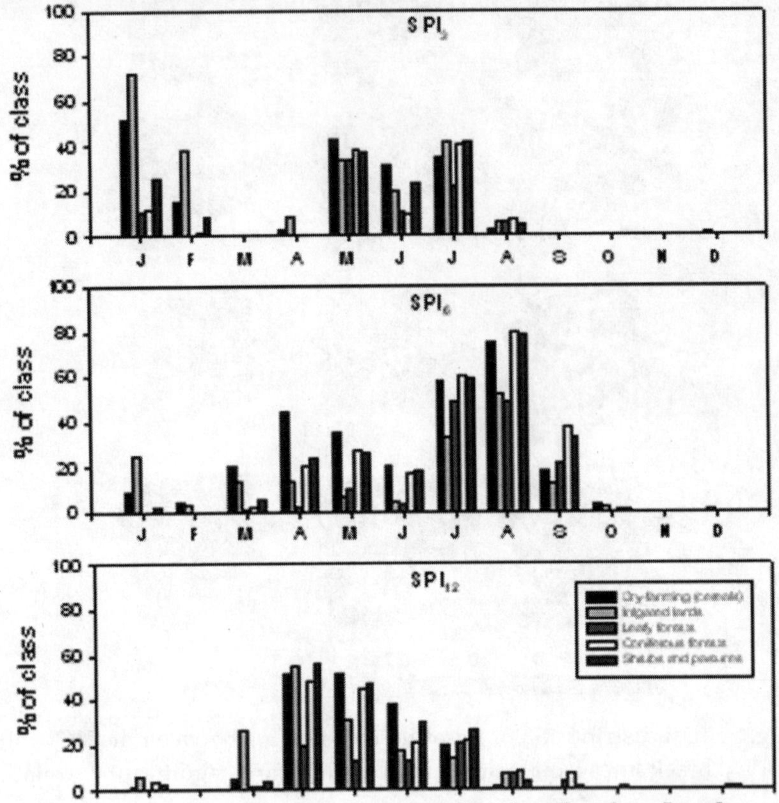

Fig. 10 per cent of class of each land-cover type in which there are significant correlations ($R = 0.53$, $p < 0.05$) between the SPI and the VCI.

The correlations between the average VCI and the 3-, 6- and 12-month SPIs for each land-cover type and month. The correlation values were obtained from the average of the whole pixels corresponding to each land-cover type. Therefore, there is a smoothing of the concrete SPI-VCI relationships that reduces the R-values as a consequence of the spatial diversity. For this reason, in order to highlight the observed relationships, when $p < 0.1$, the significant R-values are also shown, thus allowing consideration of a more relaxed threshold. For SPI3, positive and significant correlations were recorded in dry-farming areas in January and also between May and July. For all other land-cover types, the number of significant correlations ($p < 0.1$) was smaller. All correlations were negative during the autumn. This behaviour could be determined by two different processes as a function of the late summer/early autumn (September-October), when the soil is likely to be dry and the activity in the areas of natural vegetation is low when the annual vegetation cycles are ending, and late autumn/early winter (October-November) when the low air temperatures will limit growth of vegetation and the water availability would not play a major role.

Table. Monthly Correlations between the SPI and the VCI. Significant and Positive Correlations are Indicates: * (R = 0.53; p < 0.1), ** (R = 0.53; p < 0.05)

	SPI3											
	J	F	M	A	My	J	Jl	Aug	S	O	N	D
Dry-farming (cereal)	0.51*	0.41	0.10	0.19	0.49*	0.48*	0.48*	0.29	-0.25	-0.13	-0.42	0.03
Irrigated lands	0.59**	0.47	0.12	0.27	0.45	0.42	0.49*	0.29	-0.25	-0.19	-0.47	0.10
Deciduous forests	0.18	0.19	-0.03	0.22	0.47	0.33	0.39	0.24	-0.17	-0.37	-0.64	-0.34
Coniferous forests	0.26	0.26	-0.03	0.20	0.49*	0.37	0.47	0.34	-0.17	-0.31	-0.62	-0.35
Shrubs and pastures	0.36	0.33	0.04	0.21	0.48*	0.43	0.49*	0.34	-0.19	-0.27	-0.56	-0.25
	SPI6											
	J	F	M	A	My	J	Jl	Aug	S	O	N	D
Dry-farming (cereal)	0.39	0.30	0.36	0.46	0.45	0.41	0.54*	0.58**	0.41	0.18	-0.24	0.01
Irrigated lands	0.44	0.33	0.34	0.28	0.27	0.25	0.45	0.51*	0.27	0.09	-0.33	0.10
Deciduous forests	-0.02	-0.07	0.00	0.25	0.35	0.20	0.50**	0.52*	0.40	-0.03	-0.45	-0.37
Coniferous forests	0.06	0.05	0.10	0.32	0.42	0.30	0.56**	0.62**	0.48*	0.07	-0.34	-0.32
Shrubs and pastures	0.19	0.15	0.20	0.37	0.43	0.35	0.54*	0.61**	0.45	0.09	-0.31	-0.23
	SPI12											
	J	F	M	A	My	J	Jl	Aug	S	O	N	D
Dry-farming (cereal)	0.25	0.12	0.29	0.51*	0.53*	0.49*	0.41	0.30	0.11	0.08	-0.13	0.00
Irrigated lands	0.33	0.21	0.41	0.52*	0.45	0.38	0.37	0.32	0.06	0.08	-0.15	0.05
Deciduous forests	0.13	0.15	0.26	0.43	0.39	0.26	0.33	0.25	0.16	0.07	-0.09	0.00
Coniferous forests	0.21	0.18	0.29	0.51*	0.51*	0.37	0.38	0.33	0.25	0.15	0.03	0.11
Shrubs and pastures	0.26	0.18	0.31	0.54*	0.52*	0.43	0.41	0.32	0.20	0.11	-0.02	0.07

For SPI6, significant correlations were recorded in April and July, these were highest in coniferous forests and shrubs and pasture-lands. Finally, for SPI12, the highest correlations were recorded in April and May, with marked differences found between land-cover types. The highest correlations were recorded in the dry-farming areas and shrubs and pasture-lands.

Therefore, marked seasonal differences can be found between the SPI-VCI correlations as a function of the time scales of SPI and the land-cover type. The strongest correlations were recorded in the dry-farming areas, the coniferous forests and in the shrubs and pasture-lands.

The VCI-SPI3,6,12 relationships in two different dry-farming areas. The relationships in the north of the study area (La Hoya de Huesca), where annual mean precipitation is 600 mm (the sample is obtained by means the average of 12 pixels - 12 km2 - of dry-farming in this area). The relationship between the VCI and the SPI was not strong and correlations were non significant in all cases (R = 0.30, R= 0.36 and R = 0.03 for the 3-, 6- and 12-month SPIs, respectively).

By contrast, in Belchite, a region in the centre of the valley, where annual precipitation is 320 mm (the sample is obtained by means the average of 18 pixels—18 km2—of dry-farming in this area), the relationships between the

VCI and the SPI, independent of time scale, were very strong. Significant correlations being recorded between the VCI and the SPI at the different time scales (R = 0.68, R= 0.59 and R = 0.69 for the 3-, 6- and 12-month SPIs, respectively).

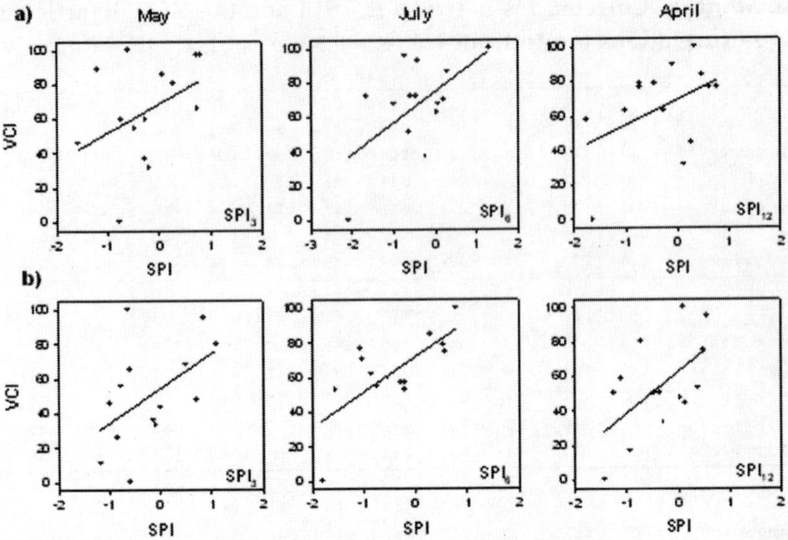

Fig. Relationships between the VCI and the SPI at different time scales (3, 6 and 12 months in May, July and April respectively) in two dry-farming areas located: a) in the North of the study area (600 mm of mean annual precipitation) and b) in the centre of the Ebro valley (320 mm of mean annual precipitation).

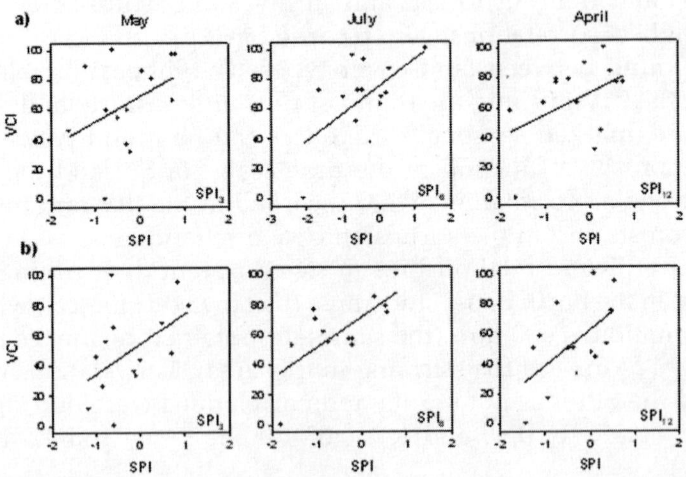

Fig. Relationships between the VCI and the SPI at different time scales (3, 6 and 12 months in May, July and April respectively) in two areas of coniferous forests located: a) in the North of the study area (850 mm of mean annual precipitation) and b) in the centre of the Ebro valley (550 mm of mean annual precipitation).

A similar behaviour was recorded when comparing two coniferous forest areas, although with lower differences between dry and humid regions in relation to the dry-farming areas. One of the forests is located in the north (Pyrenees) (12a), in an area with an annual precipitation of 850 mm, and the other in the centre of the valley (Sierra de Alcubierre) (12b), with a mean precipitation of 550 mm.

In the first of these forest areas, the only significant correlations ($p < 0.05$) were recorded between the VCI and the SPI at the time scale of 6 months (R = 0.40, R = 0.58 and R = 0.46 for the 3-, 6- and 12-month SPIs, respectively). By contrast, the coniferous forests in the centre of the valley presented higher correlations between the VCI and the SPI on the different time scales (R=0.50, R=0.73 and R=0.58 for 3-, 6- and 12- month SPIs, respectively).

In general, the R-values are higher for all land-covers for the period between April and August. This is the period when the water stored in the soil for several months (e.g., summarised by means of the SPI12) is actively transpirated by the vegetation.

In the Ebro valley, a number of experimental studies have shown that water availability during the germination period is a significant determinant of cereal development (Martí, 1992; Alberto and Machín, 1978; Austin et al., 1998), which would explain the high correlations recorded during this period (April) between the VCI and the SPI.

The Influence of Vegetation Characteristics and Climatic Conditions

The spatial distributions of the temporally averaged NDVI, CoV (Coefficient of Variation of the NDVI) and the Aridity Index (AI) were analysed in order to identify the environmental variables that determine the spatial differences in the VCI-SPI correlations. For this purpose, the April data were selected, since the highest annual vegetation activity in most of the land-cover types was recorded during this month. Moreover, SPI12 was selected for analysis since, first, it best reflects wet and dry conditions throughout the year and, second, this time scale had the most marked effects on the VCI in April.

The relationships between the SPI12-VCI correlations and the average NDVI, the CoV and the AI. Irrigated lands were not considered for analysis. The correlation between the spatial distribution of correlations between the VCI and SPI12 and the average NDVI was R = -0.54.

The correlation with the spatial distribution of CoV was R = 0.39 and, finally, the correlation between the spatial distribution of correlations between VCI and SPI12 and the spatial distribution of the Aridity Index was R = -0.46.

The high number of records analysed (18365 pixels) makes difficult to know if these correlations are statistically significant because the degrees of freedom are very high.

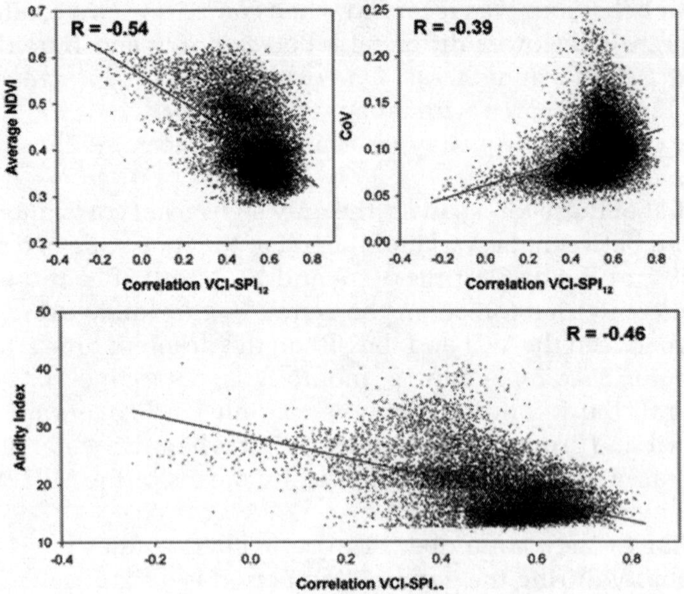

Fig. Relationships between the spatial distribution of the correlations between the VCI and the SPI12, the average NDVI, CoV and the Aridity Index (AI). The 4805 irrigation pixels were not included in this analysis, 18165 pixels remain.

For this purpose the total data base was divided in 908 random sub-samples of 20 pixels and correlations between the SPI12-VCI relationship, NDVI, CoV and AI calculated for each sub-sample. The average R values obtained for each sub-sample coincides with those values obtained for the complete dataset, and significant values (considering 20 values, R = 0.44, p <0.05) are: 65.4 per cent, 28.1 per cent and 64.0 per cent of correlations for the NDVI, CoV and AI, respectively. Therefore, the average R-values (from sub-samples of 20 values) are only significant for the NDVI and AI, whereas the correlation between the spatial distribution of SPI12-VCI correlations and the CoV is non-significant. If irrigated lands are included in the analysis, the correlations fall noticeably: R = -0.39, R = 0.27 and R = -0.36, for NDVI, CoV and AI, respectively.

Therefore, considering the whole study area, the correlation between the SPI and the VCI is, in general, higher in areas with a low vegetation cover/activity (low average NDVI values). This could indicate that areas that have a low vegetation cover/activity are more prone to the effects of drought than areas with a higher vegetation cover.

However, the opposite is the case if we compare the spatial distribution of the VCI-SPI12 correlations and the CoV values. Thus, areas that record a high interannual variability of NDVI values are more affected by drought than areas with a low interannual variability of the NDVI, although in this case the correlation is not-significant. Finally, the relationships between the spatial distribution of the AI and the VCI-SPI12 correlations indicate that the most arid areas are also more prone to the effects of drought than are the humid regions.

Therefore, taking the study area as a whole, the regions characterised by high aridity, low vegetation cover and high CoV values are those where drought has the greatest impact on the interannual variability in vegetation activity. The most humid regions, characterised by higher vegetation cover and a low interannual variability in their NDVI values, are less prone to drought.

Many of the ecosystems with low vegetation cover are located in areas with high aridity. In the steppe areas of the centre of the valley, composed of herbaceous plants and shrubs, the vegetation is adapted to deal with regular dry conditions, and has a herbaceous component in the seed-bank that could expand quickly when water resources become available. Thus, the higher correlation between the VCI and the SPI12 found in areas of low average NDVI values and high aridity could also be interpreted as a greater ability to expand the activity and the leaf area index during the humid periods and, therefore, also related to higher water use efficiency by arid ecosystems.

To assess this, the ratio between the average annual NDVI and the average annual precipitation was calculated for the different land-uses. In general, the total vegetation productivity for unity of water is a good measurement of the water use efficiency of the ecosystems to produce vegetation biomass. The NDVI has been widely used for this purpose (e.g., Nicholson et al., 1990; Tucker et al., 1991).

Land cover	**1- Total**	**2- AI<17**	**3- AI>17**
Dry-farming (cereal)	9.5	9.9	8.8
Irrigated lands	11.4	11.4	10.5
Decidious forests	8.3	10.0	8.2
Coniferous forests	9.3	10.2	8.9
Shrubs and pastures	9.3	10.2	8.5

Table. Average values of the ratio between the average annual NDVI (1987-2000) and the annual precipitation. The column 1 indicates the ratio for the whole study area, and the columns 2 and 3 show the ratio for the dry (AI < 17) and humid areas (AI > 17).

The results show that there are no important differences between the different land cover types. For the whole study area (column 1), the higher efficiency is recorded in the irrigated lands (11.4) as a consequence of irrigation. However, between the remaining land covers, where the vegetation activity depends directly on precipitation, the differences are not important. If the analysis is made considering independently the dry (AI < 17 -median value for the whole study area-) and humid areas (AI > 17), a higher water use efficiency is recorded in the driest areas but no important differences are observed in relation to the most humid regions neither between the forests and the steppe areas. Therefore, the higher correlations between NDVI and SPI12 found in the arid and low vegetated areas would not be related to a great ability to respond to the most humid periods. This is confirmed when the SPI12-VCI relationships are shown in the shrubs and pastures located in the centre of the valley (most arid regions, AI < 14).

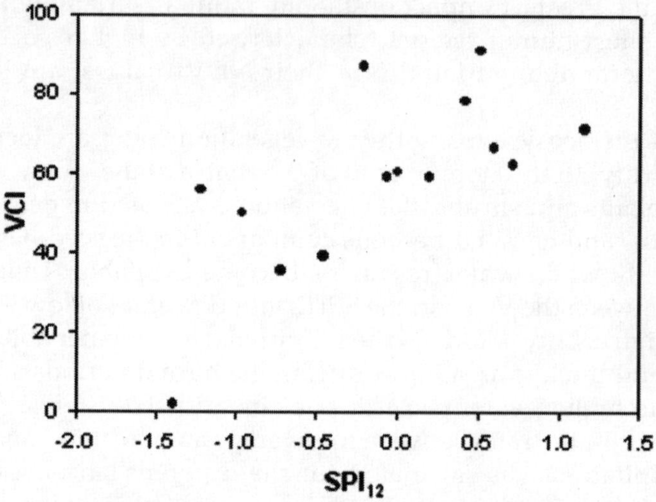

Fig. Relationship between the SPI_{12} and the VCI in the shrubs and pastures located in the most arid areas (AI < 14)

The relationship between the SPI_{12} and the VCI in these areas between 1987 and 2000. Low VCI values correspond to SPI's below 0. On the contrary, there are no differences in the VCI between the years with normal conditions (SPI ˜ 0) and the humid years (SPI > 0.5). Thus, a non-sensitivity of the vegetation activity to the SPI values higher than 0 could be inferred, which reinforces the fact that the highest SPI12-VCI correlations found in the most arid lands are mainly determined by the negative response to the drought periods. The results obtained might be affected by the spatial distribution of the land-cover types. For this reason the spatial differences in the SPI12-VCI correlations as a function of the AI, NDVI and CoV were also analysed independently in the main land-cover types of the study area.

The correlation between the spatial distribution of the SPI12-VCI correlations and the average April NDVI for different land-cover types are generally negative. However, the magnitude of these correlations varied noticeably between different land-cover types. In the irrigated lands the correlation was non-significant ($p < 0.05$). More robust negative correlations were recorded in the dry-farming areas ($R = -0.55$, $p < 0.05$), deciduous forests ($R = -0.70$, $p < 0.05$) and shrub and pasture-lands ($R = -0.46$, $p < 0.05$). The correlation in the coniferous forests was lower and non-significant ($R = -0.29$). As a consequence of the large data base the signification of correlations was evaluated by means of the average R-values from random sub-samples of 20 values, as explained below.

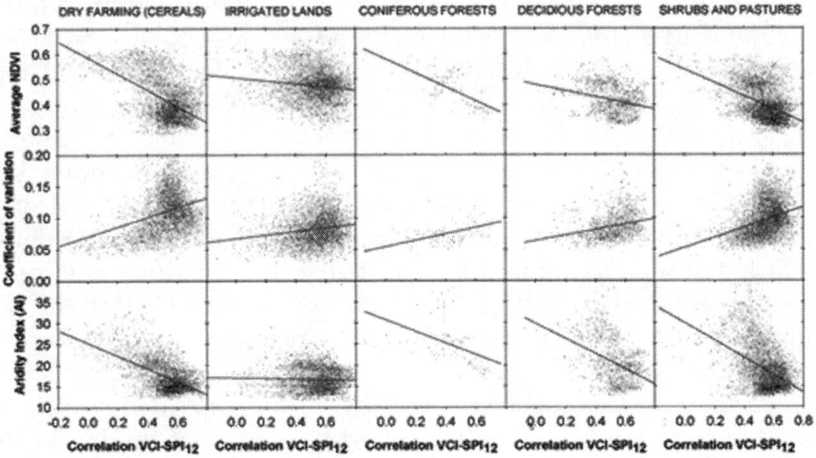

Fig. Relationships between the spatial distribution of the correlations between the VCI and the SPI_{12} and the NDVI, CoV and Aridity Index (AI) in April for the different land-cover types.

The relationships between the spatial distribution of correlations between the SPI_{12} and the VCI and the spatial distribution of the April CoV for different land-cover types were lower than those recorded for the average NDVI. The lowest correlation was also found in the irrigated lands ($R = 0.17$). In the dry-farming and the shrub and pasture lands the correlations were higher ($R = 0.30$ and $R = 0.36$, respectively) but not statistically significant. Only in the areas of deciduous forests were the correlations significant ($R = 0.49$, $p < 0.05$).

Finally, the relationships between the SPI_{12}-VCI correlations and the spatial distribution of the Aridity Index (AI) were higher than those between NDVI and CoV values and significant differences were found as a function of land-cover. The highest correlation was recorded in the dry-farming areas ($R = -0.59$, $p < 0.05$). In areas of natural vegetation, correlations were lower, and they were also significant in the coniferous forests ($R = -0.46$, $p < 0.05$) and shrub and pasture-lands ($R = -0.46$, $p < 0.05$). In the deciduous forests the correlation was non-significant ($R = -0.41$, $p < 0.05$), while similarly in the irrigated lands there was no relationship between the variables.

DISCUSSION AND CONCLUSIONS

The complexity of the drought phenomenon hinders our full understanding of their impact. This paper has shown that the effects of drought on vegetation can be highly diverse, varying with different factors including the month, land-cover type, vegetation characteristics (described by means of the NDVI), topoclimatic conditions, and the time scale of the episode. Moreover, there may be other important factors not included in this study that also can affect the spatial differences of the influence of drought on vegetation. In the middle Ebro valley (the northernmost semi-arid region in Europe), droughts have a notable impact on vegetation activity. Large areas present significant correlations between the VCI series obtained from AVHRR images and the SPI series. However, this impact is not homogeneous in space or time.

In the middle Ebro valley, the main impact of drought on the interannual differences in vegetation activity occurs in spring and summer. In these seasons, vegetation activity is highest and the role of water availability is most influential. By contrast, the cold temperatures and the frequent periods of frost recorded in winter and autumn limit vegetation activity and, therefore, although droughts occur during these periods, the role of drought is not as influential as that of low air temperature. Marked seasonal differences in the impact of drought on vegetation have been recorded elsewhere. Ji and Peters (2003) have shown that on the Great Plains of the USA the main impact of drought on vegetation activity—quantified by means of the NDVI—occurs during the period of greatest vegetation activity (spring-summer), while lower correlations are recorded at the beginning and end of the vegetation growth season.

Differences in the VCI-SPI correlations have been found as a function of the SPI time scales. In general, during the germination period in cereals, shrubs and pastures (months when a high vegetation activity is recorded), the 12-month time scale shows the highest correlation with the VCI in large areas. This indicates that vegetation activity during this critical period is mainly determined by the precipitation that has fallen over the last 12 months. The patterns and values of correlations are very similar to those obtained at the time scale of 6 months, which indicates that the winter precipitation would have the main role in explaining the interannual differences of VCI.

Studies that have analysed the precipitation-NDVI relationship show that the correlations between the two parameters are higher when considering the cumulative precipitation over a period of just several months (e.g., Wang et al., 2001; Davenport and Nicholson, 1993). Wang et al. (2003), for example, have indicated that the dry-farming and pasture productions on the Great Plains of the USA are not only determined by the precipitation recorded during the year but also by the precipitation that fell in the previous year. These studies showed that the levels of soil moisture (determined by previous year precipitation) at the start of the cultivation cycle are the main factor in accounting for the subsequent NDVI. These initial levels of moisture at the beginning of the growing season, which condition vegetation activity, are best summarised by means of a wide time scale for the drought index (i.e., SPI12).

Spatial differences in drought impact on vegetation activity are closely related to land-cover type and to topoclimatic conditions. In general, dry-farming, shrubs and pasture-lands are more prone to the effects of drought than are irrigated lands and the deciduous forests, although some areas of coniferous forest can also suffer the impact.

The highest correlations between the VCI and the SPI have been recorded in dry-farming areas. Bennie and Hensley (2001) have pointed out that, in regions with annual precipitation below 600 mm, farms require a greater surface area than those in humid areas to ensure adequate cereal production because of the greater risk of soil degradation and the lower productivity of crops. A similar behaviour has been reported in the shrub and pasture-lands of the centre of the Ebro valley, areas that have undergone major human modifications. Even during years of normal rainfall, these vegetation communities have difficulties in developing their activity and in advancing towards more mature stages. Thus, during drought episodes, these communities suffer the effects of water shortages with even greater intensity. Differences in the correlation coefficient values have also been recorded within the same land-cover type as a function of the average NDVI and climatic aridity. The impact of climate variability and drought on vegetation activity is most marked in the most arid areas. In semi-arid regions, vegetation responds rapidly to spatio-temporal variations in soil moisture (Le Houerou, 1984; Bonifacio et al., 1993; Sannier and Taylor, 1998). In general, vegetation located on the limits of its environmental distribution is more vulnerable to climatic variability than that which is located in areas of adequate climate conditions for vegetation activity.

To understand the role of precipitations on NDVI, it should also be taken into account that in humid regions the high precipitations cannot be used in its entirety by the vegetation. Therefore, there is a threshold of precipitation that increases the values of the drought index but does not necessarily increase the vegetation activity. Thus, the higher precipitations recorded in the north and south of the study area could be indicative of a lower sensitivity of the vegetation in relation to the variability of water resources. The NDVI saturation in relation to precipitation has been widely documented (e.g., Davenport and Nicholson, 1993; Malo and Nicholson, 1990; Santos and Negrín, 1997), indicating that the NDVI increases according to the precipitation until a threshold is reached. Above this threshold of precipitation, the NDVI would be non-sensitive.

Also, some physiological aspects must be taken into account to explain the strong differences found in the SPI-VCI relationships. The lower VCI-SPI correlations found in the forests in relation to the dry-farming areas and the shrubs and pastures could be related to the deeper root systems of the forests, which would diminish the effect of the water shortages of short duration. Orwing and Abrams (1997) and Jonsson et al. (2002), among others, have shown that the forests respond better to the drought indices that record the low-frequency variability of water resources. This indicates that the forests can be non-sensitive to short droughts but the longest and most intense droughts can affect the physiological structure and thus take longer to recover their normal activity. On the contrary, in the semi-arid regions, the dominant herbaceous and shrub cover develops shallow root systems which do not allow plants to reduce the negative impact of prolonged water shortages. In the study area, Castro-Díez and Montserrat-Martí (1998) have indicated that the root systems of the common shrubs and herbaceous species of the centre of the Ebro valley (e.g., Thymus s.p., Rosmarinus officinalis, Lonicera implexa) are superficial systems of a depth of less than 1 meter.

The high level of dependence of the vegetation cover in arid regions on water availability has been reported elsewhere. Abrams et al. (1998) have analysed the effects of drought on US forests. These authors have reported marked spatial variations in the response of forests to drought as a function of average climate conditions, indicating that forests located in the driest regions are more sensitive to drought occurrence. Nicholson et al. (1990) have also shown, by means of remote sensing, that the humid forests of East Africa show a low interannual variability in their NDVIs, even though annual precipitation fluctuates noticeably. However, in shrubs and pasture-lands the same authors have reported a significant response of the NDVI to annual precipitation. In Brazil, Santos and Negrín (1997) have also shown a stronger relationship between precipitation and NDVI in dry areas than in humid regions.

In the middle Ebro valley, the soil formations in the most arid areas may also play an important role in exacerbating the effects of drought on vegetation. In the centre of the valley, the soils are predominantly composed of limestone and gypsums which are poor retainers of water. Moreover, these soils are very fragile, exhibiting general land degradation and specific soil erosion, and this may in part prevent the development of a high amount of vegetation cover.

This paper has demonstrated, in line with other studies (Wilhelmi and Wilhite, 2002; Wu and Wilhite, 2004), that the impact of vegetation on climate variability, including drought, can vary spatially. In some ecosystems this behaviour could not be related to a higher vulnerability to drought. The steppes (composed of shrubs and pastures) located in the areas regularly impacted by dry conditions, have adapted strategies to deal with climate variability and droughts. Thus, the decrease in the vegetation activity in relation to drought is a normal behaviour of these ecosystems. Nevertheless, although this hydro-ecological equilibrium can be identified in some ecosystems, the vegetation cover in the middle Ebro valley has also been modified for centuries by humans. Large areas have a vegetation cover not well adapted to climate variability, in which drought noticeably affects the vegetation's variability, as Creus and Saz (2004) have shown in the reforestations located in the arid areas of the centre of the valley. Also, in the introduced crops, the high role of droughts on the VCI variability recorded in the driest areas is related to higher economic losses for farmers and, thus, to a higher socioeconomic vulnerability to drought. Moreover, although the vegetation of the steppe areas is well adapted to the drought variability, a higher vulnerability could be inferred to the economic activities that use the pastures of the driest regions, such as the extensive livestock based on sheep flocks, which are very important for the socio-economy of the region (Errea and Lasanta, 1993). Sheep raising would suffer important economic losses during the dry years as a consequence of the low pasture availability and the need to contribute with complementary food to maintain the livestock.

The monthly maps showing the correlation between the VCI and the SPI are empirical analyses of ecologic and/or economic vulnerability (a high positive correlation indicates greater ecologic or economic vulnerability to drought). These analyses complement other spatial approaches based on agroclimatological studies, while adding information regarding natural vegetation vulnerability and seasonal differences on the effects of drought. The use of such maps should improve drought management plans and play a large role in mitigating the impact of such episodes.

10

Drought and Sustainable Agriculture

BACKGROUND

Floodplain agriculture is critical to survival and economic development in the rural areas of Sahel West Africa. Semi-arid Northern Nigeria, like the entire West African Sahel, is experiencing significant climatic and anthropogenic changes that have exacerbated the food insecurity in the region rendering poor rural households more vulnerable. Following widespread failure of large-scale irrigation projects throughout much of Sub-Saharan Africa, many policy makers, stakeholders and donor agencies have become convinced that the security offered by floodplains could become the basis for rural economic transformation and food security in semi-arid Africa, as they contribute significantly to domestic food production.

Whereas, rain-fed agriculture lasts about three months, floodplain (fadama) agriculture is practiced throughout the dry season, which may last from six to nine months.

The floodplains support residual moisture agriculture at the end of the rainy season and small-scale irrigation during the dry season, thus providing opportunities for agricultural diversification not found in the uplands, and allowing for double, or even triple cropping during one year.

During the past two decades, however, most floodplains in semi-arid northern Nigeria have come under severe pressure from several sources. First, rainfall has been declining in the area especially since after the "jump point" which lies after the pre-drought period before 1964 and during the drought period between 1964 and 1972.

This change, which increased the vulnerability of rain-fed agriculture, has caused many farmers to shift their attention to fadama agriculture. Declining rainfall has also led to water scarcity in the more arid regions, which were used by nomadic herdsmen for grazing their herds during the dry season. Now, these herdsmen routinely spend the entire dry season in the semi-arid region, bringing added pressure and competition on fadama resources, principally, alluvial aquifers but also the surrounding land in the area. Secondly, widespread acknowledgement that the Green Revolution Initiative of the 1970's failed to ensure food security in Sub-Saharan Africa, compelled policy makers to shift attention away from the large-scale government irrigation projects to fadama agriculture. For example, in Northern Nigeria, funding by the World Bank facilitated the drilling of thousands of tube wells in floodplains and the distribution of petrol-driven irrigation pumps at subsidized prices to farmers to stimulate adoption of irrigated agriculture. Such large-scale investment in small-scale, floodplain irrigation sanctioned by the World Bank, illustrates the extent of the expectation that policy makers have come to place on alluvial aquifers. But these investments have increased pressure on the floodplains. This pressure coupled with the possibility of an increase in drought occurrences as projected by the Third Assessment Report of the Intergovernmental Panel on Climate Change (IPCC), could spell disaster for the sustainability of floodplain agriculture and food security in the region.

STUDY AREA

The study area is the Hadejia River valley stretching from the Kano area through Hadejia and Nguru to Gashua and covering an area of about 4,000 ha of farmland. About 8,000 farmers from different village areas are engaged in some form of fadama farming along this river valley which forms part of the Hadejia-Jama'are – Yobe River Basin. The entire Basin covers an area of 45,000 km2 and is situated in semi-arid Northern Nigeria. The three main rivers in the Basin are the Hadejia, Jama'are and Yobe rivers. The Hadejia and Jama'are meet in the Hadejia-Nguru Wetlands to become the Yobe River. The traditional farming system in the Basin, particularly in the downstream areas is flood farming. This flood farming depends on residual moisture for the second crop while rainfed farming is relied upon for the growth of the first crop. Since the early 1980s, small-scale irrigation which pumps water from the river and floodplain (shallow wells) has been stimulated in the basin through subsidies, permitting the introduction of the third crop during the dry season, while intensifying the cultivation of the second. A combination of these activities has increased the productivity and income of farmers in the area.

The choice of the study area is hinged on three main factors. First, background data on important hydro-geologic variables (e.g. stream flow stage and discharge, water level measurements, and meteorological data) for the floodplain exist from prior research conducted in the area by the researcher and other collaborators. Second, rapport has already been established and maintained with the local inhabitants since 1994 and particularly, 2003; this will be critical in assuring participation by villagers in project design, data collection, and monitoring. Third, the entire Hadejia-Jama'are – Yobe River Basin is a very important economic hub, with about 800,000 fadama farmers engaged in all year round crop production. The all-year round agricultural production, besides providing steady income to the rural populace, encourages the retention and attraction of productive labour force into the rural areas, thereby enhancing productivity and development in the rural areas as well as checking rural-urban migration with all its attendant problems.

The adverse impact of climate change in the economic sustainability of this area, is therefore of great importance. For example, the Third Assessment Report of the Intergovernmental Panel on Climate Change emphasised the need for the assessment of the impact of climate change on food and agricultural production.

For the case study, two areas were selected – Bunkure and Hadejia areas. Two villages from each area were then selected. In Bunkure area, Zangon Buhari and Barkum villages were selected. Zongon Buhari has an 'organized system' that uses surface water released from a barrage (dam) in a network of canals for irrigation. Contrary, Barkum has a 'traditional system' that uses surface and groundwater from natural sources (tributary streams) for irrigation. Similarly, in Hadejia area, Matsa and Gamsarka-Zimoni villages were selected. Matsa has a 'traditional system' while Gamsarka-Zimoni, an 'organized system'. These two farming systems are considered here as two distinct household typologies. Consequently, the two household types (traditional and organized) formed the basis of the study (unit of enquiry). These subdivisions are necessary in other to make comparisons between household types and between locations. Varying the household types according to farming system will reveal the level of vulnerability of each category to farm inputs, especially water availability, especially during and after drought events.

OBJECTIVES

The aim of this study is to investigate the vulnerability of floodplain agriculture to the potential effects of future climate change, particularly droughts, and how the farmers can adapt to such climatic hazards. The study will focus on the risks that fadama farmers are exposed to under current climate conditions, the way they adapt to such risks, and the possible variations in these adaptations that might arise in the face of global environmental change, especially further decline in rainfall leading to droughts, reduction in stream flow, lowering of groundwater levels, and increased water scarcity for livestock and humans. Understanding the relationships that have evolved to cope with current climate hazards is a critical first step in identifying possible adaptations to manage future climate risks. Therefore, the study will analyse current vulnerability and evaluate current adaptation strategies to determine their present and potential future effectiveness in reducing this vulnerability. This aim can be achieved through the following specific objectives:

- To determine how farmers are experiencing the impacts of drought in the floodplains;

- To document farmers' past and current coping strategies to climate variability and change and to evaluate the success of the current measures that have been adopted to cope with drought;
- To determine which of the coping strategies or measures fadama farmers are willing to adopt in an event of future droughts in the light of global environmental change.

METHODOLOGY

The objectives of the present study will be met by documenting past and current coping strategies and analysing the current coping strategies of farming households in the floodplain area in the face of increasing incidence of drought. The household is the main unit of analyses. The analysis will compare the two household typologies (traditional and organized) among the four villages and between the two study areas.

The method of study involved the stages enumerated below:

- An extensive literature search to conceptualize the study and to develop appropriate survey and analytical tools;
- A "transect walk" in the study area that facilitated the selection of the two sample villages from the two study areas (sites). The sample villages were selected based on the following criteria: (i) a concentration of fadama farmers within a fairly closed section of the basin; (ii) existence of organized and traditional systems with well demarcated farm plots; (iii) their proximity to sources of secondary data; and (iv) ease in communication with farmers and accessibility to the farm plots;
- A reconnaissance survey during which arrangements were made for data collection including visits to sources of information, selection of keys informants (community/religious leaders, household heads, a few farmers, Local Government representatives, and other stakeholders), development of the survey instrument (questionnaire), and selection of samples from two farming systems (i) The traditional irrigation system in which local farmers use stream (surface) water, direct from source and groundwater pumped with water pumping machines to irrigate their crops and (ii) The organized irrigation systems in which local farmers are provided with surfaced water from a barrage (reservoir) released through a main canal, distributory and field channels;
- Field data collection (questionnaire survey and focus group discussions – FGDs). The field survey was carried out with the farming household as the unit of analysis. The questionnaire was the major tool for data collection. The questionnaire designed for data collection is divided into six sections as follows:
  - Identification and demographic characteristics;

- Educational attainment and employment (income generation economic activities);
- Drought history and environmental concerns (vulnerability);
- Impacts of droughts (and water availability) on Fadama (floodplain) agricultural activities;
- Impacts of droughts (and water availability) on household livelihood; and
- Coping strategies and Adaptation.

The questionnaires were administered to a sample of fifty (50) farmers randomly selected from each of the four communities (villages). This gives a total of 200 questionnaires in all. Effort was made to include farmers in different age groups (20 to 49 years and 50 years and older). Both male and female respondents were considered but it was not possible to get a gender balance because only a few females participated as a result Islamic religion restrictions on women (the purdah system). The focus group discussions (FGD) with 10 to 12 members were conducted for the two areas separately, irrespective of farming system (household type). The FGD, were conducted by the researcher (as facilitator) with the two earlier research assistants taking the notes;

- Data analysis at this stage only involved the documentation, evaluation and description of farmer responses to the questions asked in the questionnaire as well as comments generated during the focus group discussions (FGD). This was done using descriptive statistics including frequencies, percentages and averages (means) and presented on tables and graphs (figures). Furthermore, the several coping strategies identified from the questionnaire administration and focus group discussions were analyzed to ascertain appropriate adaptation measure against future droughts based on farmer knowledge and preferences, using a participatory rural appraisal (PRA) technique, Walker, et al, (2002), Roncerel, et al, (2003), and Downing (2004)).

DEMOGRAPHIC AND SOCIOECONOMIC CHARACTERISTIC OF FADAMA FARMERS

DEMOGRAPHIC CHARACTERISTICS

In Zangon Buhari, there were 47 males and only 3 females (females are not usually allowed to meet with strangers because of the religious restriction called pudah). Respondents were between the ages of 25 years and 93 years. 48 per cent were less that 50 years and 52 per cent of them were 50 years and above. All but one of the respondents was married. The unmarried respondent is an elderly woman who is now a widow. Majority (46 per cent) of those married had two wives. The highest category of respondents (34 per cent) had between 5 and 9 children. Majority (62 per cent) of them had between one and four other dependants. The highest category (34 per cent) of respondents had household sizes of between 10 and 14 members. This is closely followed by those who had between 5 and 9 members (28 per cent). From the 50 households interviewed, there are 650 members with an average household size of 13 people.

In Barkum, There were also 47 males and only 3 females between the ages of 20 years and 80 years. 44 per cent were less that 50 years and 56 per cent of them were 50 years and above. Almost all the respondents were married except the elderly women who participated that are widows. Majority (60 per cent) of the married respondents had two wives. The highest category of respondents (30 per cent) had between 10 and 14 children. The highest category (38 per cent) of respondents had between one and four dependants. The highest category (30 per cent) of respondents had household sizes of between 15 and 19 members. This is closely followed by those who had between 10 and 14 members (26 per cent). From the 50 households interviewed, there are 968 members with an average household size of about 19 people.

All respondents in Matsa were males between the ages of 20 years and 80 years. 60 per cent were less that 50 years and 40 per cent of them were 50 years and above. Almost all the respondents were married but one. Majority (54 per cent) of the married respondents had two wives. This is followed by those with one wife (36 per cent). The highest category of respondents (44 per cent) had between 5 and 9 children. The highest category (38 per cent) of respondents had between one and four dependants. This is closely followed by those (34 per cent) who did not have dependants. The highest category (38 per cent) of respondents had household sizes of between 10 and 14 members. This is closely followed by those who had between 15 and 19 members (20 per cent). From the 50 households interviewed, there are 819 members with an average household size about 16 people.

While in Gamsarka-Zimoni, all respondents were males between the ages of 21 years and 70 years. 54 per cent were less that 50 years and 46 per cent of them were 50 years and above. All the respondents here were married, majority (56 per cent) of whom had two wives. This is followed by those with one wife (32 per cent). The highest category of respondents (44 per cent) had between 5 and 9 children. The highest category (42 per cent) of respondents had between one and four dependants. This is followed by those (26 per cent) who had ten or more dependants. The highest category (28 per cent) of respondents had household sizes of between 10 and 14 members. This is closely followed by those who had between 5 and 9 members (26 per cent). From the 50 households interviewed, there are 713 members with an average household size about 14 people. Barkum had the highest household size followed by Matsa, Gamsarka-Zimoni and lastly Zangon Buhari.

Educational Status of Respondents

Most (86 per cent) of the respondents in Zangon Buhari had Koranic education, one had no formal education, four with primary and two with secondary education. Most those in Barkum (72 per cent) had Koranic education, four had no formal education, eight with primary and two with secondary education. For the respondents in Matsa, most of them (76 per cent) of the respondents had Koranic education, one had no formal education, ten with primary education and one with secondary education. Whereas those in Gamsarka-Zimoni, a little less than half (48 per cent) had Koranic education, only eight had no formal education, 15 with primary education and three with secondary education.

Major Source of Income

Generally speaking, agriculture (farming and animal husbandry) is the major economic activity for most people in the study area. For Zangon Buhari, the major source of income for a majority of the respondents (72 per cent) is farming. Few others are traders (8 per cent), while the remainder provide off-farm labour or do some crafts as artisans (20 per cent). Most (78 per cent) of the farmers are engaged in a combination of rainfed and irrigated farming, producing the first crop and third crop respectively. (The second crop is cultivated at the end of the rainy season using residual moisture and when the first is maturing and awaiting harvesting). Only two (4 per cent) practice irrigation farming alone and nine others (18 per cent) farm only in the wet seasonal.

Those engaged in rainfed and irrigated agriculture, work during both seasons and spend between eight to twelve months of the year farming. The others work during any of the seasons and spend only about four months of the year working off-farm. Most of the farmers (38 per cent) earned between N50,000 and N99,000 from farming at the end of the season. This is followed by those who earn between N10,000 to N49,000 in a season, accounting for 22 per cent. A reasonable percentage (14 per cent) of the respondents earn more than N200,000 at the end of the farming season. Farmers in this community earn as much as N 4,233,600 at the end of the farming season. This gives an average income of about N84,600 per household.

Similarly, in Barkum, the major source of income for a majority of the respondents (74 per cent) is farming. A few others are traders (4 per cent), while the remainder provide farm labour or do some crafts as artisans (22 per cent). Almost all (90 per cent) of the farmers are engaged in a combination of rainfed and irrigated farming. Only two of them (4 per cent) practice irrigation farming alone and three (6 per cent) farm only in the wet seasonal. Farmers earned reasonable income from farming. The highest category of the farmers (26 per cent) earned up to N200,000 and more at the end of the farming season. This is followed by those who earn between N10,000 to N49,000 in a season, accounting for 20 per cent. A reasonable percentage (16 per cent) of the respondents earn between N50,000 and N99,000 at the end of the farming season. Farmers in this community earned as much as N7,045,450 at the end of the farming season with an average household income of about N140,900.

In Matsa also, farming is the major source of income for most of the respondents (78 per cent). A few others are traders (12 per cent), while the remainder provide farm labour or do some crafts as artisans (10 per cent). Almost all (88 per cent) of the farmers are engaged in rainfed and irrigated farming. Only two (6 per cent) of the farmers practice irrigation farming alone while three (6 per cent) farm only in the wet season. These earned the highest category of the farmers (32 per cent) between N10,000 and N49,000 at the end of the farming season. This is followed by those who earned between N50,000 to N99,000 in a season, accounting for 22 per cent. A reasonable percentage (16 per cent) of the respondents earn between N100,000 and N149,000. Farmers in this community earn as much as N 4,189,000 at the end of the farming season and gave an average income of about N83,700 per household.

While in Gamsarka-Zimoni, even a greater majority of the respondents (88 per cent) have farming as their major source of income. Very few are traders (2 per cent), with the remainder engaged in farm labour and off-farm activities (craftsmen and artisans) accounting for only 10 per cent. Virtually all (98 per cent) of the farmers are engaged in rainfed and irrigated farming. Only one (2 per cent) practice irrigation farming alone and none farm only in the wet seasonal. Their major economic activity earned them a lot of income. The highest category of the farmers (48 per cent) earned N200,000 or more at the end of the farming season. The next higher category earned between N100,000 to N149,000 in a season, accounting for 18 per cent. A reasonable percentage (16 per cent) of the respondents earn between N50,000 and N99,000. In total, farmers in this community earned as much as N 10,201,000 with an average income of about N204,000 per household. This high income can be attributed to contribution of five farmers who earned large sums. Four earned as much as half a million and one earned up to one million.

Other Sources of Income

Most respondents from the four villages have additional (secondary) sources of income from activities undertaken during the wet and/or dry season. In Zangon Buhari, majority (62 per cent) do some kind of trading and other off-farm jobs for additional income from which they generated some income. A majority of the respondents (45.6 per cent) earned between N10,000 and N49,000, followed by those who earned less than N10,000. Only two persons (2.5 per cent) earned more than N200,000 from secondary sources of income.

Similarly, majority of respondents (62 per cent) in Barkum do some kind of trading and other off-farm jobs for additional income from which most of them (about 46 per cent) earned between N10,000 and N49,000. This is followed by those who earned less than N10,000. Only two persons (2.5 per cent) earned more than N200,000 from secondary sources of income.

On the other hand, a higher percentage of the respondents (68 per cent) in Matsa are engaged in trading and other off-farm jobs for additional income. These earned them between N10,000 and N49,000 for a majority of the respondents (about 54 per cent). The next category earned between N50,000 and N99,000. But as in the case of Barkum, only two persons (1.4 per cent) earned more than N200,000 from secondary sources of income.

Lastly, respondents in Gamsarka-Zimoni also earned income from secondary activities. However, a lower percentage of the respondents (54 per cent), compared to the other three villages trade and do other off-farm jobs for additional income. Consequently, less income is generated from these sources. A small majority of respondents (39.7 per cent) earned between N10,000 and N49,000 from these other sources. This is followed by those who earned between N50,000 and N99,000.00 (22.1 per cent).

INCIDENCES OF DROUGHT AND IMPACTS ON FADAMA FARMING

Farmers' experience of drought

All respondents from the four communities agreed that they have observed changes in rainfall. However, the nature of observation and experience vary from one community to another. In Zangon Buhari, almost all of the respondents (98 per cent) indicated that they have observed a decrease. However, only one person representing only 2 per cent of respondents indicated an increase.

These changes were observed within the last three decades with the highest frequency being in 1984, followed by 1985. These dates coincide with the drought period of 1984/85. In Barkum, most of the respondents (88 per cent) indicated that they have observed a decrease. However, six persons representing 12 per cent of respondents indicated an increase in rainfall in the area. These changes were observed within the last four decades (1964 to 2004) with the highest frequency being in 1985, followed by 2003. These dates also coincide with the drought period of 1984/85 and another in 2003/2004. In Matsa, most of the respondents (88 per cent) indicated that they have observed a decrease as the case in Barkum. However, six persons representing 12 per cent of respondents indicated an increase in rainfall. These changes were observed within the last four decades with the highest frequency being in 2004, followed by 1985, then 2003. These dates also coincide with the drought period of 1984/85 and another in 2003/2004.

Finally, in Gamsarka-Zimoni, most of the respondents (80 per cent) indicated that they have observed a decrease in rainfall. However, six persons representing 20 per cent of respondents indicated an increase in rainfall. These changes were observed within the last four decades (1963 to 2004) with the highest frequency being in 2004. This date coincides with the drought period of 2003/2004.

Table. Number of Respondents who have Observed Changes in

	Zangon Buhari		Barkum		Matsa		Gamsarka-Zimoni	
Changes observed	Decrease	Increase	Decrease	Increase	Decrease	Increase	Decrease	Increase
Total rainfall	49	1	44	6	44	6	40	6
Time of onset	28	19	26	20	23	8	34	6
Time of cessation	29	16	20	23	7	20	7	33
Frequency of events	34	7	42	3	27	6	40	3
Intensity of events	31	8	41	4	29	2	40	1
Level of groundwater	17	31	30	17	21	9	11	11
Timing of recharge	27	17	31	10	21	9	14	9
Timing of discharge	25	19	18	23	5	25	7	16
Level of stream flow	32	8	36	11	18	16	13	29
Recharge of stream flow	34	6	40	5	10	16	10	13
Discharge of stream flow	32	8	19	25	10	15	17	9
Frequency of floods	34	4	37	5	8	28	11	28
Total number of observations	372	144	384	152	223	160	244	164
Percentage (%)	72.1	27.9	71.6	28.4	58.2	41.8	59.8	40.2

From the table, it can be observed that in Zangon Buhari, There was a decrease in water for all the observations made, except for the level of groundwater. This shows a general decline in water availability, which coincided with the drought periods of 1984/85 and 2002/2003 noted by the respondents.

In Barkum, There was also a general decrease in water for all the observations made, except for time of cessation of rainfall, as well as timing of discharge of groundwater and stream flow. This also shows a general decline which coincided with the drought periods of 1984/85 and 2002/2003. In Matsa, again, there was a general decrease in water for all the observations made, except for time of cessation of rainfall and timing of discharge of groundwater.

There was also an increase in recharge and discharge of stream flow and frequency of floods. This general decline in water coincided with the drought periods of 1984/85 and 2002/2003/2004. While in Gamsarka-Zimoni, there was a general decrease in water for all the observations made, except for time of cessation of rainfall, timing of discharge of groundwater, level of stream flow, recharge of stream flow, and frequency of floods. This general decline in water coincided with the drought period 2002/2003/2004.

Whereas in the more southerly locations, Zangon Buhari and Barkum decreases in water availability with less frequencies of floods were observed, in the more northerly locations, Matsa and Gamsarka-Zimoni, although there were decreases in water availability, there were increases in the incidences of floods.

However, this was perhaps superseded with increased frequencies in flooding. This may be attributed to the fact that these two locations are at the downstream section of the River Hadejia. These fluctuations between water shortages and excesses certainly affected farming activities in the area. If such conditions prevail in the future, fadama farming will surely be affected due to droughts and flood insurgences.

Other environmental problems affecting fadama farming

Farmers were asked to rank environmental problems as they affect their farming activities. The outcomes are summarized on Table below.

Table. Ranking of Environmental Problems in the Study Area

Environmental Problem	No of respondents that rank each problem											
	Zangon Buhari			Barkum			Matsa			Gamsarka-Zimoni		
	1st	2nd	3rd	1st	2nd	3rd	1st	2nd	3rd	1st	2nd	3rd
Decline in rainfall	19	6	2	34	7	1	36	9	2	37	5	1
Dust storms	2	3	3	1	3	5	2	2	8	1	8	7
Desertification	1	3	9	0	3	4	3	1	1	1	5	2
Deforestation	5	4	3	0	4	1	1	1	5	0	0	3
Soil erosion	3	4	6	2	3	11	0	4	6	0	2	3
Decline in soil moisture	4	11	6	2	8	12	0	3	9	3	4	11
Declining soil fertility	4	6	5	4	11	7	6	26	4	2	19	1
Loss of biodiversity	0	1	4	1	3	2	1	0	2	0	1	14
Total	**38**	**38**	**38**	**44**	**42**	**43**	**49**	**46**	**37**	**44**	**44**	**42**

The table demonstrates an assessment of respondents' ranking of environmental problems. Whichever environmental problem has more respondents in any rank category, it is selected as the dominant rank in that category.

In Zangon Buhari, declining rainfall is ranked 1st by the majority. Therefore, declining rainfall is considered as the most serious problem. This is followed by decline in soil moisture (which is drought related) and then desertification in a descending order of severity. In Barkum, Therefore, declining rainfall is considered as the most serious problem followed by decline in soil fertility and declining soil moisture. These rankings are similar to those of Zangon Buhari. In Matsa, declining rainfall is the most serious problem.

This is followed by declining soil fertility, then declining soil moisture. Lastly, in Gamsarka-Zimoni, declining rainfall is also the most severe environmental problem. This is followed by declining soil fertility and declining soil moisture.

From all villages, declining rainfall ranked 1st meaning that it is a common problem in the Hadejia river floodplain. This is confirmed by the ranking of other problems which are quite similar among the four villages.

DEFINITION OF DROUGHT FROM FARMERS' PERSPECTIVES

Farmers' definition of drought also followed a similar pattern as seen on Table.

Table. Farmers' Definition of Drought

Definition	Zangon Buhari Freq	Zangon Buhari %	Barkum Freq	Barkum %	Matsa Freq	Matsa %	Gamsarka-Zimoni Freq	Gamsarka-Zimoni %
Lack of rain	6	12	26	52	13	26	34	68
Shortage of rain during rainy season	24	48	13	26	10	20	11	22
Lack of water	3	6	3	6	18	36	1	2
Cessation of rain during rainy season	10	20	5	10	1	2	1	2
When rainfall starts and stops midway and crops dries due to lack of water	5	10	2	4	5	10	2	4
Lack of both rain and underground water	2	4	1	2	3	6	1	2
Total	**50**	**100**	**50**	**100**	**50**	**100**	**50**	**100**

From Table above, farmers' definitions were concentrated as follows: In Zangon Buhari, almost half of the farmers (48 per cent) defined drought as shortage of rain during the rainy season followed by those (20 per cent) that defined it as cessation of rain during the rainy season. In Barkum, most of the farmers (52 per cent) defined drought as lack of rain followed by those (26 per cent) that defined it as shortage of rain during rainy season. In Matsa, most of the farmers (36 per cent) defined drought as lack of water followed by those (26 per cent) that defined it as lack of rain. Yet a reasonable percentage (20 per cent) that defined it as shortage of rain during rainy season. While in Gamsarka-Zimoni, Most of the farmers (68 per cent) defined drought as lack of rain followed by those (22 per cent) that defined it as shortage of rain during the rainy season.

From these definitions, it is almost certain that farmers have a fair conceptualization or understanding of what drought is and its effects.

Vulnerability to drought

Farmers were also asked to indicate the level of dependence or vulnerability of their activities to drought. Their responses and observations.

Table. Farmers' Vulnerability to Drought

Extent	Zangon Buhari Freq	Zangon Buhari %	Barkum Freq	Barkum %	Matsa Freq	Matsa %	Gamsarka-Zimoni Freq	Gamsarka-Zimoni %
Entirely dependent	7	14	17	34	33	66	35	70
Partially dependent	19	38	26	52	3	6	8	16
Neutral (Not sure)	20	40	5	10	6	12	3	6
Very slightly dependent	1	2	1	2	5	10	4	8
Not dependent at all	3	6	1	2	3	6	0	0
Total	**50**	**100**	**50**	**100**	**50**	**100**	**50**	**100**

In Zangon Buhari, most farmers (40 per cent) in this area were neutral about their vulnerability to drought. This is closely followed by those (38 per cent) who believed that they are partially dependent (vulnerable) to drought. This may be attributed to the fact that they are located more southerly or because they rely on water released from the barrage (dam). In Barkum, most farmers (52 per cent) believed that they are partially vulnerability to drought followed by those (34 per cent) who believed that they are entirely dependent (vulnerable) to drought. This may be attributed to the fact that Barkum is located further north than Zangon Buhari and the people of Barkum rely on natural sources of water (stream flow and groundwater).

In Matsa, majority of farmers (66 per cent) in this area believed that they are entirely dependent (vulnerability) to drought. This is followed by those (12 per cent) that are neutral about the dependence of their activity to drought. This may also be attributed to the fact that they are located further north than Zangon Buhari and even Barkum. Secondly, fadama farmers in this area are also dependant on natural sources of water (stream flow and groundwater). While in Gamsarka-Zimoni, majority of farmers (70 per cent) in this area believed that they are entirely dependent (vulnerability) to drought followed by those (16 per cent) that thought their activities are partially dependent on drought. This may also be attributed to the fact that Gamsarka-Zimoni are located further north as Matsa, although the fadama farmers in this area are also dependant on water supplied from the barrage (dam) through canals.

However, all the farmers (from the four villages) agreed that water is necessary for farming, especially during droughts which they have all experienced. A majority of the farmers (44 per cent) in Zangon Buhari experienced the drought in 1984. Other droughts they experienced were in1976, 1982, 1984, 1985, 1986, 2002, 2003, and 2004 and most of them (58 per cent) also agreed that the most devastating drought was in 1984. In Barkum, farmers were more diverse in their experience of drought. Majority of them (18 per cent) experienced the drought of 1984, followed by those of 2000 and 2003 (16 per cent each), then 1985 (14 per cent). A greater majority (30 per cent) also agreed that the most devastating drought was in 1984. This is closely followed by those (28 per cent) who believe that the drought of 1985 was the most devastating.

Similarly, in Matsa, a majority of the farmers (30 per cent) experienced the drought in 2004, followed by those the experienced it in 1985 (16 per cent), then 1983 (14 per cent). A greater majority (36 per cent) also agreed that the most devastating drought was in 1985. Although this does not conform to the majority that experienced the 2004 drought, it conforms to the next, 1985. While in Gamsarka-Zimoni, the story is fairly different. Most of the farmers (70 per cent) experienced the drought in 2004, followed by those the experienced it in 1985 (6 per cent).

Albeit, a smaller majority of the farmers (24 per cent) agreed that the most devastating drought was in 2004. This is closely followed by those that believed the 1983 drought was devastating (20 per cent), then 1985 (18 per cent). These observations conformed to the majority that believed the 2004 drought was the most devastating.

From the figure above, it is obvious that the devastating effects of drought vary from one community to the other. For example, in Zangon Buhari, where farmers rely on surface water channelled from a barrage (dam), almost all the farmers (92 per cent) indicated that the most devastating effect of the drought was that their crops dried up and were destroyed. But in Matsa, a community that depends on both surface water and groundwater, the effects are fairly (almost evenly) distributed.

However, for all communities, the most devastating effect of droughts experiences by fadama farmers was that their crops dried up and destroyed. The next effect that cut across all communities is food scarcity and starvation. Reduction in quantity of food intake (including skipping of meals) was less intense in all villages.

COPING WITH DROUGHT

Coping with drought depends on the severity of the stressors or level of vulnerability and the resilience or decisions and/or actions farmers may take to minimise the impacts. The myriad of decisions taken by farmers were documented and are discussed in the subsequent sections.

Major decisions taken to cope with drought

Farmers from the different villages took a variety of decisions to cope with the most devastating drought. From the figure, it can be observed that in Zangon Buhari, majority (52 per cent) of farmers sold their assets in order to cope with the drought. This is followed by those (28 per cent) who intensified irrigation (fadama farming) in order to cope. In Barkum, farmers were moderate in their decisions. Majority of the farmers (28 per cent) intensified irrigation (fadama farming), followed by those who migrated (24 per cent), then those that embarked on off-farm labour and trade (20 per cent) in order to cope. In Matsa, half (50 per cent) of farmers migrated (in search of jobs, money and food) in order to cope with the drought. This is followed by those who intensified irrigation (20 per cent) and those who embarked on off-farm labour and trade (16 per cent). While in Gamsarka-Zimoni, more than half (58 per cent) of farmers intensified irrigation in order to cope with the drought, followed by those who migrated (22 per cent). A reasonable proportion of the farmers embarked on off-farm labour and trade (16 per cent) in order to cope.

It can be seen that all four villages had different major strategies to cope with drought. However, irrigation was the first in Barkum but second in the other three villages. This signifies the importance of irrigated (fadama) farming in the Hadejia River Valley. The main reasons for their decisions are the need to feed their families for those that intensified irrigation or sold assets and to look (search) for money and food for those that migrated or embarked on off-farm labour.

Almost all farmers from all four villages (81 per cent, 100 per cent, 86 per cent, and 84 per cent of farmers in Zangon Buhari, Barkum, Matsa, and Gamsarka-Zimoni respectively) agreed that the will make decisions based on the same reasons, should another drought occur.

ASSISTANCE AND ADVICE FOR FARMERS DURING AND AFTER DROUGHT

Farmers expect assistance from family members, neighbours, the community, government, and non-governmental organizations. The basic assistance they require from these bodies is food but in some cases, fertilizer (and other farm inputs), money and prayer. They also require advice as to what to due after a drought event. They seem not to be familiar with the need to plan against a future occurrence. Fewer farmers had ever received advice in order to cope with a drought or prepare against another. Figure shows the extent of lack of information dissemination in the area.

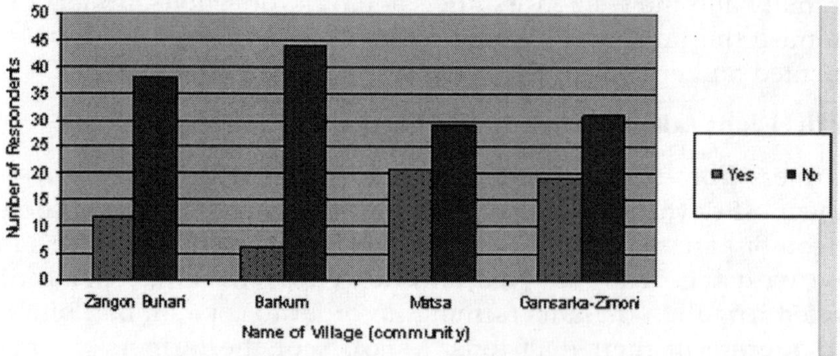

Receipt of Assistance or Advice by Farmers

From the figure, it is obvious that majority of farmers received neither assistance nor advice. For example, in Zangon Buhari and Barkum in Bunkure area of Kano State, very few farmers received any advice or assistance from any agency on how to prepare for or cope from a drought. The few received advice from the Kano State Agricultural Development Authority (KNADA), Kano State Corporative (KASCO) and government extension workers. The assistance offered included most especially food (grains), clothes, shelter, money, and farm inputs. These advice and assistance were given during the drought years, 1984 to 1988, especially in 1984 in Zangon Buhari; as well as some of the drought years, 1975, 1985, 1987, 1998, 2000 and 2002 in Barkum.

Matsa and Gamsarka-Zimoni in Hadejia area, Jigawa State had faired better in term is receipt of advice and assistance although majority did not receive any. For those that received, the assistance came from the government organizations including Jigawa Agricultural and Rural Development Authority (JARDA), the radio and extension workers as well as community development associations. Most of the assistance and advice were offered very recently (2000, 2001, 2002, 2004 and 2005) and included food stuff, farm inputs, cash loans and a water pump.

Existing Plans in an Event a Drought Occurs very Soon

As far as existing plans in an event of another drought, there is variation among the four villages. Most of the farmers had existing plans to cope in the future in an event of another drought. A few farmers, however, did not have any existing plan for coping with drought in the near future but to pray. Such attitude may be attributed to their level of education or lack of access to information (e.g. warning signs). Figure below illustrates some of the variations among farmers in respect to future plans to cope with drought.

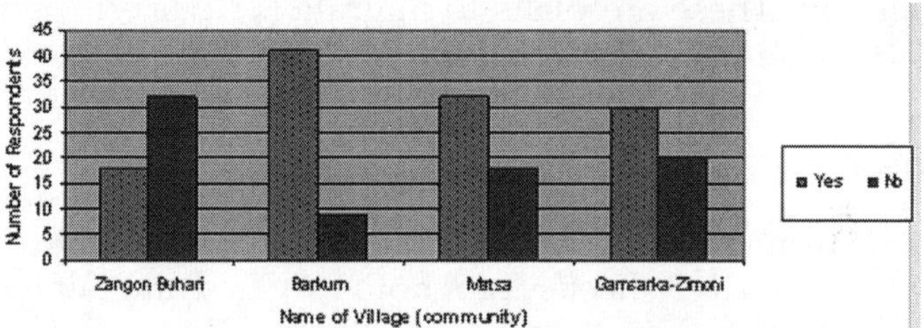

Existing Plans in an Event of another Drought

In Zangon Buhari, most farmers (64 per cent) have no existing plans. Only a few (36 per cent) have existing plans. But for such group of farmers, prayer was their main reason or plan in an event of another drought. In Barkum, however, the opposite is the case since majority of the farmers here (82 per cent) have existing plans. Only 18 per cent of them have no existing plans. There is a similar trend in Matsa and Gamsarka-Zimoni with 64 per cent and 60 per cent of farmers respectively, having existing plans.

Major Strategies to Cope with Future Droughts

For the group of farmers who had existing plans, such plans are considered as coping strategies. The major strategies farmers plan to adopt in an event of another drought soon.

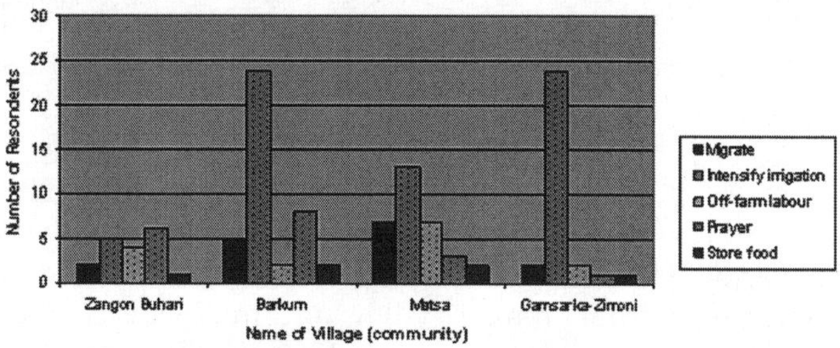

Major Strategies to Cope with Future Drought

From the graph, it can be observed that the existing plans or coping strategies vary from one village to the other. In Zangon Buhari for example, fewer farmers had existing plans. Majority of the farmers (33 per cent) plan to commit the problem to prayers in an event of another drought. This is closely followed by those that plan to intensify irrigation (28 per cent) and embark on off-farm labour (22 per cent). In Barkum however, more farmers have planned coping strategies. Majority of them (59 per cent) plan to intensify irrigation in an event of another drought. This is followed by those (20 per cent) that plan to commit the problem to prayers. This is a departure from the first village where fewer farmers have plans.

Matsa and Gamsarka-Zimoni have similar plans, with intensification of irrigation topping the list in each case with 41 per cent and 80 per cent respectively. Both villages have corresponding strategies tailing plans to intensify irrigation. These are plans to migrate (22 per cent) and to embark on off-farm labour (22 per cent) for Matsa 7 per cent for each case in Gamsarka-Zimoni.

Impacts of Droughts on Fadama Agricultural Activities

Majority of farmers in the Hadejia river valley own land ranging from 0.2 hectares to as much as 35 hectares for both irrigated and rainfed agriculture. A few may rent land from others or use community land. Some portion of this land is dedicated for irrigation (fadama) farming, especially land located along the floodplains of stream channels (tributaries) of the Hadejia River. Land located along the dam canals are particularly used for irrigation. The reason why farmers may dedicate land or want to intensify irrigation farming in the floodplains (fadama) will depend on access to surface or groundwater as well as the the impacts of drought on their activities. Table below presents data on land dedicated for fadama farming in the study villages.

Table. Area Extent of Land under Irrigation (Fadama) Farming

	Zangon Buhari		Barkum		Matsa		Gamsarka-Zimoni	
Farm size (ha)	Freq	%	Freq	%	Freq	%	Freq	%
Less than 1	4	8	18	36	13	26	2	4
1 - 1.99	12	24	8	16	5	10	4	8
2 - 2.99	14	28	8	16	7	14	1	2
3 - 3.99	6	12	5	10	4	8	2	4
4 - 4.99	9	18	2	4	4	8	2	4
5 - 5.99	3	6	7	14	3	6	5	10
6+	2	4	2	4	14	28	34	68
Total	**50**	**100**	**50**	**100**	**50**	**100**	**50**	**100**

In Zangon Buhari, majority of the farmers (20 per cent) have farm sizes of between 3.00 and 3.99 ha. This is followed by those with 2.00 to 2.99 ha (14 per cent). In Barkum, farmers have smaller land masses with the majority (36 per cent) having farm sizes of less than one hectare. This is followed by those with farm sizes ranging from 1 to 1.99 ha and 2.00 to 2.99 ha (16 per cent each). In Matsa, farmers dedicated larger pieces of land for irrigation. Majority of the farmers (28 per cent) here have farm sizes of six hectares or more. This is closely followed by those with farm sizes of less than one hectare (26 per cent), then 2 to 2.99 ha (14 per cent). While in Gamsarka-Zimoni, just as in Matsa, farmers dedicated large pieces of land for irrigation. Most of the farmers (68 per cent) here have farm sizes of six hectares or more. This is followed by those with farm sizes between 5 to 5.9 hectares (10 per cent), then between 1 and 1.99 (8 per cent).

Sources of Water for Fadama Farming

Fadama farmers obtain water from different sources for farming. Some of these sources include: rain harvesting, flood water (residual moisture), surface water (direct from stream or river, direct from pond, dam or barrage) and groundwater (hand dug wells {up-land}, shallow wells {river bed}, and boreholes or tube wells).

Almost all of the farmers (98 per cent) in Zangon Buhari used surface water from the dam via canals mostly during the rainy season. The only farmer (2 per cent) that does not use water from the canal got water from shallow wells dung in the farm (used in-situ). In addition, 56 per cent of them used rainwater harvesting in the rainy season for their farms. No farmers in Zangon Buhari used water direct from a stream, residual moisture or borehole (tube well).

In Barkum, majority of the farmers (86 per cent) used only groundwater by means of tube wells mostly during the rainy season. Only 14 per cent of the farmers use groundwater from shallow wells, hand dug wells or draw water direct from the stream using water pumping machines. In addition to groundwater sources, only nine farmers (18 per cent) used rainwater harvesting techniques in the rainy season for their farms. One other farmer used residual moisture at the end of the rainy season to grow his crops (this is the second crop). No farmer in this area has access to water from a dam conveyed in canals unlike Zangon Buhari area.

In Matsa, most of the farmers (43 per cent) used groundwater from either shallow wells, borehole or hand dug wells mostly during the dry season. This is closely followed by those farmers (40 per cent) that used surface water abstracted directly from the stream or ponds usually at the end of the rainy season. Water abstraction by both groups of farmers is done by means of petrol-driven water pumping machines. In addition to groundwater sources, only seven farmers (10 per cent) use rainwater harvesting techniques in the rainy season for their farms in this area. Five other farmers (about 7 per cent) use residual moisture at the end of the rainy season to grow a second crop. However, no farmer in this area has access to water from a dam conveyed in canals.

While as many as 80 per cent of the farmers in Gamsarka-Zimoni used only surface water drawn by gravity from the canals fed by the barrage (dam) mostly during the dry season. Only 10 per cent of the farmers used groundwater from shallow wells, hand dug wells or borehole in this area usually during the rainy season. In addition to groundwater sources, a few farmers (about 10 per cent) used rainwater harvesting techniques in the rainy season for their farms. No farmer here used residual moisture at the end of the rainy season to grow his crops.

FUTURE ADAPTATION STRATEGIES (MEASURES)

The coping strategies farmers have adopted in the past and present will obviously become adaptation strategies for the fadama farmers during future droughts. In Zangon Buhari, the five most widely acceptable strategies farmers are willing to adopt in the future are the use of early maturing crops, irrigated crops, culling animals, food storage, and keeping of livestock. Replanting of crops when affected by a drought is the sixth most important strategy fadama farmers are willing to adopt in the future. The most important strategies farmers are not willing to adopt in the future are planting of drought resistant crop varieties, crop diversification, use of high yield and low input crop varieties, keeping of livestock and selling of assets.

In Barkum, the five most widely acceptable strategies farmers are willing to adopt in the future are early maturing crops, irrigated crops, drought resistant crop varieties, livestock diversification, and crop diversification. Crop diversification has displaced food storage in this case which comes as the sixth strategy farmers are willing to adopt. The most important strategies farmers are not willing to adopt in the future are to abandon their activity, migrate, or cultivate drought resistant varieties. In Matsa on the other hand, the five most widely acceptable strategies farmers are willing to adopt in the future are irrigated crops, early maturing crops, low input varieties, drought resistant crop varieties, and crop diversification. The most important strategies farmers are not willing to adopt in the future are migrate, culling animals, selling assets, change farm location, get assistance from NGOs.

Whereas in Gamsarka-Zimoni, the five most widely acceptable strategies farmers are willing to adopt in the future are irrigated crops, early maturing crops, drought resistant crop varieties, crop diversification, and food storage. The most important strategies farmers are not willing to adopt in the future are crop diversification, culling animals, get assistance from relations, assistance from community among others.

Farmers that are not willing to adopt scientific strategies have exhibited their level of vulnerability involved with some of these strategies. This may be attributed to the lack information, extension services or the required inputs to grow the crops. In fact, some of the scientific (biogenetic) crops have high requirements to thrive well (produce good yields). However, those that are more willing to adopt scientific strategies may be less vulnerable. This may be attributed to the fact that they have access to advice and information from extension services or government. Sometimes they are provided with farm inputs to grow the crops. In fact, some of the scientific (biogenetic) crops have high requirements to thrive well (produce good yields).

Choice of Appropriate Future Adaptation Strategies (Measures)

Farmers were given the option to choose from a number of strategies they will adopt in the future in an event of another drought.

SUSTAINING AGRICULTURE IN THE ERA OF CLIMATE CHANGE IN INDIA

The imperative to shift to sustainable farming is more urgent than ever in the era of anthropogenic climate change, though the imperatives for such a shift flow from other factors too (including increasing enormous public financing needs for inputs like chemical fertilizers, the lack of economic viability of the current model of farming for millions of farmers, the environmental health disaster that is unfolding in different parts of the country due to intensive agricultural technologies, the ecological degradation of productive resources in these intensive models etc).

Climate change is already a reality for a majority of Indian farmers even as plans are being evolved at the government level mostly to create adaptive capabilities; meanwhile, Indian farmers are being forced to adapt to several CC-related changes by themselves because they have no other choice. For no fault of theirs, Indian farmers, like the most marginalized everywhere, are paying a high price for anthropogenic climate change. The worst-hit, as usual, are small and marginal holders in marginalized locations with social disadvantages to begin with. Such farmers have meager resources to buffer them from the new risks that climate change poses.

The macro-plans being created by the government in India are mostly meant for the National Agricultural Research System (NARS) and a re-arranging of their research agenda in the context of climate change – the NARS's linear systems of "lab to land" research have been already critiqued for their non-participatory, top-down, unaccountable nature. Such systems which are centred on unilateral knowledge generation only within agricultural universities to be then carried to the farmers for adoption through the agricultural extension system, discounting farmers' (traditional) resources and knowledge, will be grossly inadequate for the immediate needs of adaptation. The adaptation requirements are to be met urgently, with no time lost (the time taken for conventional knowledge generation by the NARS to be then 'transferred' to farmers is ill-affordable here). This is where a new learning paradigm is needed – to learn from farmer innovators, from traditional knowledge and resources and disseminating useful practices and resources to others. This position paper, while agreeing in principle with India's international stand in climate negotiations that agriculture-related Green

House Gas emissions cannot be equated in any manner with lifestyle-related GHG emissions and appreciates the 'common but differentiated action' demand, takes a stand that if sustainable agriculture is promoted and established even for adaptation reasons, it will result in mitigation of GHGs too. Sustainable Agriculture practices in farming are therefore a win-win option where mitigation cannot be interpreted as coming in the way of equitable and just growth of the nation. These practices as the following sections show, contribute to increased food and nutritional security, contribute to sustainability of productive resources and improvements in rural livelihoods. They also lead to mitigation of GHG emissions from farming.

The Indian government has no reason to shy away from understanding and stating the negative impacts of GHG-inducing intensive models that have been promoted so far. It is time that fundamental shifts are made in the agricultural technologies that are promoted in the country, to address the imperative of climate change (amongst other imperatives).

CLIMATE CHANGE AND AGRICULTURE IN INDIA

India is a large country with 15 agro-climatic zones, with diverse seasons, crops and farming systems. For a majority of people in India, to this day, agriculture is the main source of livelihood. Agriculture is the most vulnerable sector to CC as it is inherently sensitive to climate variability and CC will leave its impacts on Indian agriculture in various direct and indirect ways. This obviously means an impact on the lives and livelihoods of millions of Indians.

For instance, it is reported that about two-thirds of the sown area in the country is drought-prone and around 40 million hectares is flood-prone. The poorest people are likely to be hardest hit by the impacts of climate variability and change because they rely heavily on climate-sensitive sectors such as rainfed agriculture and fisheries. They also tend to be located geographically in more exposed or marginal areas, such as flood plains or nutrient-poor soils. The poor also are less able to respond due to limited human, institutional and financial capacity and have very limited ability to cope with climate impacts and to adapt to a changing hazard burden.

Impacts of CC on agriculture: Climate change is manifesting itself in many ways across the country. Among the indicators, while long term rainfall data analysis shows no clear trend of change, regional variations as well as increased rainfall during summer and reduced number of rainy days can be noticed. In the case of temperature, there is a 0.6o C rise in the last 100 years and it is projected to rise by 3.5-5o C by 2100. The carbon dioxide concentration is increasing by 1.9 ppm each year and is expected to reach 550 ppm by 2050 and 700 ppm by 2100. Extreme events like frequency of heat and cold waves, droughts and floods have been observed in the last decade. The sea level has risen by 2.5mm every year since 1950 while the Himalayan glaciers are retreating. These are all symptomatic of climate change. Available research indicates that climate change-induced rise in temperature is going to affect rainfall patterns – farming in India depends on monsoons and there is a close link between climate and water resources.

The organic carbon levels and moisture in the soil will go down while the incidence of runoff erosion will increase. The quality of the crop will also undergo change with lower levels of nitrogen and protein and an increased level of amylase content. In paddy, zinc and iron content will go down which will impact reproductive health of animals. Insect lifecycles will increase which in turn will raise the incidence of pest attacks and virulence. Other likely impacts are change in farm ecology viz. bird-insect relations, and an increase in the sea levels which will cause salinity ingression and submergence.

It is projected that due to climate change, kharif rainfall is going to increase and this might be positive for kharif crops. Further, for kharif crops, a one-degree rise in temperature may not have big implications for productivity. However, temperature rise in rabi season will impact production of wheat, a critical food-grain crop.

The surface air temperatures will increase by 2 to 4°C by 2070-2100. As mentioned earlier, the rabi crop will be impacted seriously and every 1oC increase in temperature reduces wheat production by 4-5 million tons, as per a study by IARI. This loss can be reduced to 1-2 million tons only if farmers change to timely planting. Increased climatic extremes like droughts and floods are likely to increase production variability. Productivity of most cereals would decrease due to increase in temperature and decrease in water availability, especially in Indo-Gangetic plains. The loss in crop production is projected at 10-40 per cent by 2100, depending upon the modeling technique applied.

The impacts of climate change are already visible. A network of 15 centres of ICAR working on studying climate change has reported that apple production is declining in Himachal Pradesh due to inadequate chilling. This is also causing a shift in the growing zone to higher elevations. Similarly, in the case of marine fisheries, it has been observed that Sardines are shifting from the Arabian Sea to the Bay of Bengal, which is not their normal habitat. In fact, fisheries are the most vulnerable sector to climate change. Crops have the ability to adapt to extreme climate variability even up to, say, 40C while fishes and animals do not. It has also been recorded that the pest ecology of certain crops is changing due to climate change.

Extreme weather conditions resulting in disasters will obviously have their own socio-economic impacts, especially on the poor. Further, changes in crop productivity will have implications on farmers' incomes. Changes in agronomic practices etc., will

Impacts of agriculture on Climate Change: While climate change affects Indian farming and farmers' livelihoods adversely, the converse is also true –

Indian agriculture, even if not in the same degree as the developed world's agriculture, does contribute to Climate Change.

Capital depletion and massive additions of external inputs in a 'linear' model (as opposed to cyclical systems in ecological farming models) are characteristic features of intensive, industrialized models of agriculture. In this model, a farm is treated like a factory with inputs and outputs calculated in a monocropped situation with grain yield given the highest importance compared to any other parameter and often, the externalized costs are ignored.

Amongst various GHGs that contribute to global warming, carbondioxide is released through agriculture by way of burning of fossil fuel; methane is emitted through agricultural practices like inundated paddy fields, for example; nitrous oxide through fertilizers, combustion of fossil fuels etc. Nitrous oxide has a global warming potential 296 times greater than CO2. In India, it is estimated that 28 per cent of the GHG emissions are from agriculture; about 78 per cent of methane and nitrous oxide emissions are also estimated to be from agriculture.

As per the Intergovernmental Panel on Climate Change (IPCC), every quintal of nitrogen applied in farming emits 1.25 kg of nitrous oxide and globally half of the nitrogen applied to crops is lost to the environment. The greenhouse gas emissions from fertiliser manufacture and use in India reached nearly 50 million tonnes of CO2-equivalent in 2006/07, which represents about 3 percent of total Indian greenhouse gas emissions.

Another major contributor of GHGs is the burning of crop residues. In Punjab, wheat crop residue from 5,500 square kilometers and paddy crop residues from 12,685 square kilometers are burnt each year. Every 4 tons of rice or wheat grain produces about 6 tons of straw. Emission Factors for wheat residue burning are estimated as: CO- 34.66g/Kg, NOx – 2.63g/Kg, CH4 – 0.41g/Km, PM10 – 3.99g/Kg, PM2.5 – 3.76g/Kg.

Burning of crop residues also impacts the soil (fertility). Heat from burning straw penetrates into the soil up to 1 cm, elevating the temperature as high as 33.8–42.2°C. Bacterial and fungal populations are decreased immediately and substantially in the top 2.5 cm of the soil upon burning. Repeated burning in the field permanently diminishes the bacterial population by more than 50 per cent. The economic loss due to the burning of crop residues is colossal. Each year 19.6 million tonnes of straw of rice and wheat, worth crores of rupees are burnt. Used as recycled biomass, this potentially translates into 38.5 lakh tonnes of organic carbon, 59,000 tonnes of nitrogen, 2,000 tonnes of phosphorous and 34,000 tonnes of potassium every year.

Another potent GHG is methane which is emitted in copious amounts through inundated paddy cultivation. In India, of a total area of 99.5 Mha under cereal cultivation, 42.3 Mha (or 42.5 per cent) is under rice cultivation. It is grown under flooded conditions and the seedbed preparation involves puddling or plowing when the soil is wet to destroy aggregates and reduce the infiltration rate of water. Such anaerobic conditions lead to emission of methane and possibly nitrous oxide through inefficient fertilizer use. Emission of methane from rice paddies in India is estimated at 2.4 to 6 Teragram (Tg) out of the world total emission of 25.4 to 54 Tg from all sources and 16 to 34 Tg from rice cultivation. The average methane flux from rice paddies ranges from 9 to 46 g/m2 over a 120- to 150-day growing season.

Another indirect contribution of agriculture to GHG emissions comes in the form of large dams. Large dams contribute 18.7 per cent of emissions in India as per an estimate. Total methane emissions from India's large dams could be 33.5 million tonnes (MT) per annum, including emissions from reservoirs (1.1 MT), spillways (13.2 MT) and turbines of hydropower dams (19.2 MT). The livestock sector is another major contributor to production of GHGs. For the year 1997, livestock contributed 9.0 Tg methane and 1 Gg nitrous oxide which in terms of CO equivalent it is around 190 Tg. Shift to stall-fed systems of livestock rearing creates problems with the dung while shift from fodder to feed, concentrates etc. carries higher ecological foot prints.

There is also the issue of more energy use in intensive farming models in the form of fossil fuels for machinery like tractors, harvesters and so on, pumps for irrigation etc.

POTENTIAL OF SUSTAINABLE AGRICULTURE AS LOW-GHG, RESILIENT FARMING SYSTEMS

Sustainable Agriculture can be defined as an integrated farming system (with crops, trees, livestock etc.) which is based on locally adapted agro-diverse cropping patterns and use of local resources (natural resources and natural processes), based on local knowledge, skills and innovations.

This Position Paper would also like to highlight the potential of sustainable agriculture, in terms of mitigation of GHG emissions as well as adaptation to climate change. Further, sustainable agriculture holds great potential for meeting global and national food security requirements even as it leads to improvement of farmers' livelihoods through enhancing their net incomes and improving the productivity of their resources in the long run.

Food security: A question that is often posed with regard to sustainable agriculture or organic farming is whether it will be able to feed the growing population. We respond by saying that sustainable agriculture does not imply lowered yields, as experience of successful farmers bears out on the ground. This is reinforced by an FAO report (2007) which says that "conversion of global agriculture to organic management, without converting wild lands to agriculture and without using N-fertilisers would result in a global agricultural supply of 2640 to 4380 Kcal/person/day". Sustainable intensification in developing countries through organic practices would increase production by 56 per cent. A meta-analysis of 133 scientific papers concluded that organic agriculture was particularly competitive under lower yield environments, a feature that is common in developing countries. Organic yields on average are comparable to conventional yields although yields do decline initially when converting from high-input systems and almost double when converting from low-input systems. In India, it should be remembered that a majority of land is rainfed and continues to be low-input by default.

Improvement in rural livelihoods: While macro-level food production and availability levels are likely to increase through sustainable agriculture, at the individual and community level too, there are bound to be improvements. As FAO notes again, access to food will increase by livelihood improvement both for farmers and agricultural workers through organic farming. Organic agriculture improves food access by increasing productivity, diversity and conservation of natural resources, by raising incomes, improving employment and by reducing risks. It has been recorded that shift to sustainable agriculture practices can reduce the outward migration from rural areas.

Reduction in GHG emissions: Changes in farming models and practices towards sustainable agriculture offer a significant opportunity at reducing GHG emissions. Organic farms use on an average 33 to 56 per cent less energy per hectare, as per FAO (2007).

Organic farming reduces its fossil fuel dependence in many ways. For instance, for soil productivity management, internal inputs and practices are used rather than chemical fertilizers – for instance, creating the micro-climate required for increased soil (beneficial) microbial activity. This is done by returning bio-mass to the soil. Legume production, crop rotation, mixed cropping etc., are other ways of achieving this. Pest management also does not depend on chemical pesticides but a variety of local resources and practices.

IFOAM notes that avoidance of methane emission is also possible through organic agriculture – through the promotion of aerobic micro-organisms and high biological activity in soils, oxidation of methane can be increased. Through practices like System of Rice Intensification, which is mostly based on principles of ecological farming, flooding in rice paddies can be reduced and thereby, methane emissions.

Nitrous oxide, result of overdoses and losses on nitrogen, can be effectively minimized through sustainable agriculture practices. While production of chemical fertilizers is an energy-intensive process that emits carbondioxide and nitrous oxide, application of nitrogen fertilizers makes the soil emit nitrous oxide. These can be avoided through organic farming.

Sustainable agriculture also increases the Soil Organic Carbon (SOC) by incorporating organic materials into the soil. Soil can be a major source of storage of carbon, about twice as much carbon as in the atmosphere. Fertiliser use replaces soil organic matter in intensive systems, which reduces potential sequestration. Crop, tree and livestock integration with a systematic recycling of organic wastes is an integral part of sustainable agriculture. Long term studies have shown that compost application and cover crops in rotation were particularly adept at increasing soil organic matter even in comparison to no-tillage techniques. While conservation tillage is promoted elsewhere as a way of sequestration of carbondioxide, this is often done by the use of chemical herbicides and GMOs which have their own ecological implications. In sustainable agriculture however, mitigation of climate change can be addressed both carbon sequestration and minimized emissions of GHGs. Agro-forestry is also a desired principle of organic farming which further adds to the potential of SA in carbon sequestration.

Creation of resilient systems leading to better adaptation:

Extreme and unpredictable weather conditions are part of the reality of climate change even as temperature rise and changes in rainfall, changes in pest and disease incidence etc., will also be the stark reality for farmers. What the situation then requires are resilient and adaptive farming systems with the least amount of loss to the productive resources, production and the farmer.

One of the most important requirements for adaptation would be farmers' knowledge, in negotiating complex agro-ecosystems. As a philosophical approach, organic farming has always laid thrust on farmers' skills, knowledge, innovation, horizontal sharing, observations and intuition etc. Several large organic farming projects across the world have built successful institutional models for systematic support for farmers' knowledge and innovation and constant enhancement. This forms a key part of the adaptation potential of sustainable agriculture.

To address extreme weather conditions, organic farms will be better suited. The better drainage and water holding capacity of organic soils reduces the risk of drought and soil erosion, for instance. Organic farming practices are in a good position to maintain productivity in the event of drought, irregular rainfall events and rising temperatures, notes a recent technical paper from International Trade Center (WTO) and FiBL. This paper notes that soils under organic management retain significantly more rainwater thanks to the "sponge properties" of organic matter. Water percolation is 15-20 per cent more in organic systems. Water capture in organic plots was twice as high as conventional plots during torrential rains, which in turn reduces the risk of floods.

The most important component of organic systems - diversity - contributes a lot to the resilience of organic farms. Enhanced biodiversity of organic farms have several positive ecological implications – pest prevention, and similar effects on diseases, better utilization of soil nutrients and water etc. Organic farming is also associated with decreased irrigation needs by about 30-50 per cent. This becomes an important part of adaptation in drought conditions. Given the mitigation and adaptation potential of organic systems, as well as the potential to increase food security, it becomes imperative to shift agriculture towards more ecological farming models which are sustainable. Potential of Organic Farming beyond purely agricultural technologies: Organic farming often also focuses on consumer behaviour and encourages lower ecological footprints through localized food production and consumption and reducing food miles too. Responsible consumer behaviour also goes a long way in mitigating emissions since it is found that a significant contribution to GHG emissions is also from long food miles and any plans for addressing CC should focus on the entire chain from farm to plate.

This paper would like to reinforce that such farming, even though farmers adopting it are at a disadvantage due to lack of support systems in the form of extension, marketing, grassroots institutions etc., is already being practiced successfully in lakhs of acres all over the country. In large government-supported sustainable agriculture projects like Community Managed Sustainable Agriculture (CMSA) in Andhra Pradesh, where women farmers are taking a lead in implementing a large ecological farming project on more than ten lakh acres, it has been found that it is possible to scale up organic farming onto large areas, with sensitive support systems built along with people's institutions at the village level. The CMSA programme, being implemented since 2005, has shown that farmers do tend to adopt and spread ecological practices at a rapid pace provided they are collectivized into village-level institutions and provided that appropriate extension support is provided from the village upwards. In this programme, available data shows that there has been no fall in yields for farmers shifting to organic farming. Data shows that the village economy stands to gain a lot due to savings on expenditure on external inputs and net incomes of individual farmers rise.

INDIA'S NATIONAL ACTION PLAN ON CLIMATE CHANGE (NAPCC)

India has announced a National Action Plan on Climate Change in August 2008. The NAPCC's formulation processes were found to be very top-down and non-participatory by many analysts.

The NAPCC proposes to address climate change-related issues in India through the setting up of eight inter-connected Missions: National Solar Mission; National Mission for Enhanced Energy Efficiency; National Mission on Sustainable Habitat; National Water Mission; National Mission for Sustaining the Himalayan Ecosystem; National Mission for a "Green India"; National Mission for Sustainable Agriculture and National Mission on Strategic Knowledge for Climate Change.

The following are some of the key points that emerged as a response from civil society to the NAPCC proposals.

- Definition required: Sustainable Agriculture is a misnomer for what has been proposed in the NAPCC, under the name of Sustainable Agriculture and therefore, a correct, common understanding of the term is required. The current set of proposals would not lead to improving the soil health, central to sustainable agriculture, nor to cyclical models of farming, internalizing farm inputs (including crop waste) into farming systems, which define sustainable agriculture. On the contrary, the existing suggestions would continue the conventional linear, intensive models that further the existing dependency of farmers on external agencies for everything, including for knowledge. That is one of the reasons for the current day crisis in agriculture.

- Creating the imperative for the paradigm shift: The NAPCC makes no mention and assessment of Green Revolution-induced climate change in India. Shying away from stating the issues with the current model of agriculture will not create the imperative for a shift to sustainable agriculture, which is a requirement both for mitigation as well as adaptation. The NAPCC should clearly specify incentives to farmers for shifting to organic farming and sustainable agriculture practices. The government should realize that the imperative to shift to sustainable agriculture is larger than climate change.

The NAPCC, especially in the sections related to agriculture, does not bring up mitigation possibilities at all – while that could be a posture adopted at the international level, for common but differentiated responses, it is interesting to note that the only place where mitigation is mentioned is to make the entry of Genetically Engineered crops a possibility. Further, the NAPCC should expressly acknowledge the potential that exists of mitigating GHG emissions from farming through a shift to organic farming.

- Policy approach: Strategies should be evolved for a time-bound phasing out of climate change-inducing practices towards sustainable agriculture with clear targets and financial outlays. This includes a focus on the role of pasture lands, fisheries, animal husbandry (rather than the bias on crop husbandry that is present in the NAPCC) and seed banks governed by farmers' bodies as major thrust areas for adaptation. Or, when plans are made about access to information, the emphasis should not be just on information packages to farmers in a top-down manner about geo-spatial impacts of climate change, but also data on conventional vs. organic practices so that informed choices can be made by farmers.
- Biotechnology: On the use of biotechnology, especially genetic engineering, as part of the NAPCC, it is felt strongly that the government should focus on reducing the present subsidies to GHG-emitting practices like fertilizers rather than come up with GE seed varieties which are supposed to reduce GHGs. In fact, an assessment of the stress (in)tolerance of GE crops should be an important part of understanding the implications of Genetic Engineering as an

agricultural technology in the era of climate change. There is no one GM commercial plant available throughout the world that has the characteristics described on p.37, although this has been claimed for advertising for the last ten years. Even on the biotech company basis, the GM plants include in 2009 63 per cent herbicide tolerant plants that can grow only with the herbicide company's treatment, 15 per cent of new insecticide producing plant making a kg/ha of pesticide, and 22 per cent of both combined characters (or 2 herbicide tolerances + 1 or 2 insecticide production in the plant). The more complex characters such as stress responses are driven by almost 50 genes that are not even cloned all today. There will be no agronomical truth behind that. Then once obtained maybe, it will take another 10-15 years to develop these crops on a worldwide basis, since todays crops have been developed 15 years ago.

- 'Land to lab' programmatic interventions: The NAPCC focuses too much on setting a research agenda for the NARS (National Agricultural Research System), following the old model of 'lab to land' research and not so much about programmes to be implemented immediately at the farmers' level. In the context of climate change and adaptation, there is hardly any time to be lost and farmers' need for resilient systems cannot wait for more research in the old paradigm to be taken up. The need is for solutions discovered from the farms, assessed and validated and spread to others, especially in terms of adaptation. There is a strong opinion that there is enough evidence of time-tested practices and experiences from the ground of certain sustainable agriculture principles and practices creating resilient farming systems. Further, civil society organizations also have enormous experience with creating effective people's institutions at the ground level which will allow for the delivery of programmes in an effective fashion. Therefore, the overwhelming need is for immediate programmatic interventions drawing on the strength of traditional knowledge and resources, farmers' innovations and experiences with the civil society.

Alternative, horizontal extension systems with farmers' organizations at the centre are an important part of information-centred addressal of climate change.

As part of the NAPCC, capacity building of agriculture scientists and extension workers on organic farming should be taken up so that they are equipped to take the message to farmers.

- Traditional knowledge and resources: The National Action Plan does not give adequate prominence to traditional resources and knowledge, which need to be made a cornerstone for interventions on sustainable agriculture. It was felt that popularization of traditional knowledge in addition to ever-evolving innovations in the fields of practicing organic farmers should be considered to be an important component of adaptation to climate change in agriculture. Such farmers should be identified and lessons learnt and disseminated through the extension system.

There should be an emphasis on falling back on indigenous resources (seeds, animal breeds etc.), which have proven track record of adaptation to stress conditions.

The Plan should also make Seeds, as replicable resources in the hands of farmers institutionalised the form of seed banks, as a major thrust and strategy for adaptation. As part of the NAPCC, there should be a mechanism evolved to track and monitor genetic erosion for all of the country due to climate change.

- Centre-State relations: State governments should be involved in consultations and planning right from the beginning – it is not enough that centrally-evolved plans are imposed upon them. In fact, it is ultimately the departments of agriculture and the extension and delivery mechanisms at the state level which will directly take everything to farmers and support them to bear the consequences of climate change. For instance, seed rolling plans need to be evolved by each state, with an emphasis on revival and restoration of open-pollinated, traditional and locally-adapted varieties.
- 'Public-People' Partnership: Similarly, it was felt that civil society and its institutions should also be involved in planning and implementation related to the NAPCC. For instance, alternative, horizontal extension systems with farmers' organizations at the centre are an important part of information dissemination and learning for adaptation to climate change. The stress should be on public-people partnership in the Plan. There should be recognition that only market-driven technologies are not the answer in the era of climate change. There is a need for renewed thrust on public research in partnership with communities.
- Risk management: When it comes to Risk Management, it should be acknowledged first that the existing risk management strategies and mechanisms have failed farmers badly. There is a need for complete recasting of the existing models and mechanisms. We need new mechanisms to assess damage and loss and better ways to deliver support - including weather insurance, livestock insurance and effective crop insurance.
- Clear convergence: The Plan should clearly spell out how it converges with other plans and missions both within the NAPCC as well as in other agencies like the Planning Commission.
- Social safety nets: As part of adaptation strategies, strong social security nets should be put in place for the rural households,

including with a provision of minimal incomes, pension, insurance etc., with special emphasis on the agriculture workers.

PROMOTE SUSTAINABLE AGRICULTURE IN THE ERA OF CLIMATE CHANGE

Sustainable Agriculture (ecological farming/organic farming/LEISA/Non Pesticidal Management/SRI etc) approaches are now acknowledged for the wide set of ecological and economic benefits that accrue to the practitioners as well as consumers of agricultural products. These approaches which are based on low external inputs are also low energy intensive and less polluting hence mitigate and help in adapting to the climate change. However, the promotion of sustainable agriculture on a large scale is often confronted about its potential as well as its practical limitations. In the last five years two large scale initiatives, NPM scalingup (Community Managed Sustainable Agriculture-CMSA) in Andhra Pradesh and SRI promotion in states of Tripura, Orissa and Tamil Nadu have brought in new learnings and broken the earlier apprehensions on scaling up such practices and their relevance on a large scale.

These successful experiences had three elements in common. First, all have made use of locally adapted resource conserving technologies. Second, in all there has been coordinated action by groups or communities at local level. Third, there have been supportive external (or non-local) government and/or non-governmental institutions working in partnership with farmers. Almost every one of the successes has been achieved despite existing policy environments which still strongly favor 'modern and established' approaches (technology and support systems) to agricultural development. Now the challenge is to create a policy environment to scaleup across the nation. This needs a newer approach in terms of capacity building, horizontal learning, newer institutional systems and newer forms of financial support to be put in place.

The programmatic support to agriculture today favour only high external input based agriculture. As a result, none of the mainstream programs provide any support for promotion of these models. This needs the recasting of program guidelines or initiating newer program to provide support to more sustainable models in agriculture which can be easily accessible to small and marginal farmers. The mission on sustainable agriculture can initiate a programmatic support to scale up sustainable agriculture with the objectives to

- Reduce the risks and vulnerabilities with uncertain weather conditions and degraded and limited natural resources in these regions, by adopting suitable cropping patterns and production practices
- Diversify the assets and income sources to sustain the livelihoods

by integrating livestock and horticulture into agriculture and promoting on-farm and off-farm employment opportunities,

- Conserve and efficiently use the available natural resources like soil and water, and promote biomass generation,
- Organise farmers into institutions which can help them to have better planning, greater control over their production, help to access resources and support, improve food security and move up in the value chain,
- Build livelihood security systems to cope up with the natural disasters like drought, floods and other climate uncertainties
  - Sustainable agriculture: Sustainable Agriculture can be defined as an integrated farming system which is based on locally adapted cropping patterns and local resource (natural resources and natural processes) use based on local knowledge, skills and innovations.

The capacity of a farming system to adapt to changing climate and weather conditions is based on its natural resource endowment and associated economic, social, cultural and conditions. The viability of these elements also constitutes the basis for sustainable agriculture, understood as agricultural production that: ensures adequacy of food production; does not harm the resource base; is economically viable; and enhances quality of life. Many climate and weather risk management strategies fit squarely into sustainable agriculture practices and can, therefore, be promoted with several of the programs and policies targeting environmentally responsible production.

Strategies to be adopted

- Changes cropping patterns and cropping systems to suit the local resource and weather conditions. Multiple/mixed cropping, intercropping systems with legume components etc.
- Ecological farming practices which can maximise the local resource use. Many of these practices are based on indigenous knowledge and focus on building soil biological productivity. Non Pesticidal Management, Organic Soil Management, Community Seed Banks, System of Rice Intensification, soil moisture management etc have already proven to be useful.
- Locally adopted crop varieties specially in saline and flood prone areas, drought prone areas, making suitable selections adopting Participatory Plant Breeding and Participatory Varietal Selection.
- Developing suitable farming systems integrating agriculture, horticulture and livestock.
- Farmers' Institutions: Organized communities have proven to be more effective in planning and managing their resources and livelihoods, lobbying for a policy change and securing their entitlements. Appropriate institutional systems for each of the purposes need be established.

- Strategy to be adopted –
  - The farmers would be organised into common interest groups federated into producer collectives. Existing institutions like Women SHGs etc would be used to initially anchoring the program.
  - These institutions would take the roles of planning, mobilising resources, organising production, and take up post harvest management and marketing activities.
  - The producer collectives will improve the collective bargaining power of the farmers, will internalize market activities like bulking, primary and secondary processing which improve the village economy.
- Food and livelihood Security: Shift to sustainable agriculture is often seen as a compromise on food security. This is mainly because food is understood as only wheat and rice, few pulses, oilseeds and vegetables. The food basket can be increased if we can expand the scope to include millets, coarse cereals, dryland fruits, uncultivated greens etc which can also bring in nutrition security. Data from National Centre for Organic Farming (NCOF), ICRISAT and CMSA have proven that crop productivity can also be maintained with organic/ecological farming. Going beyond the current food security systems like PDS and mid-day meal schemes, systems need to be established to improve livelihood security in terms of sustaining food production in the village, improving income generation opportunities to the small farmers and agriculture labor is important in the rural areas especially in rainfed regions. The frequent monsoon failures results in droughts and support systems needs to be build to help the farm families and livestock to tide over.
  - Building house hold food security systems by adopting suitable cropping patterns
  - Village level management systems for alternative models like grain banks would be appropriate and attempted.
  - Suitable off-farm and non-farm employment opportunities would be identified and promoted.
- Financial Support Systems: Currently all the financial support systems to agriculture are given only for external inputs. We need to create proper support systems for farm internalized inputs, community based infrastructure, knowledge and skill building and sharing etc.
  - Direct Subsidies to farmers rather than input subsidies
  - Integrating NREGA with sustainable agriculture so that each farmer gets 100 labor days for farming can provide ample scope in this direction.

- Explore tools like Direct Income Support which exist in many developed/developing countries need to be explored as decent living income cannot be explored only through pricing mechanism.
- Partnerships: At the district level we need build partnerships between various governmental and non governmental agencies to implement the program. At the national level we need to build an alliance of Public sector research organisations, extension agencies, departments dealing with rural livelihoods and NGOs which are working on sustainable agriculture/ organic/ natural/ ecological farming.

SUSTAINABLE DEVELOPMENT OF INDIAN AGRICULTURE

There are many things unique about the story of technological change in Indian Agriculture in sixties, seventies and eighties in India. However, my regret is that when our experience is applied in Africa or other developing countries, all the wrong lessons are learned.

This note has three parts; first deals with the current challenges in Indian agriculture. Part two deals with the historical review of the social, political and economic forces that shaped our policies. Part three deals with the issues to be faced in the nineties.

Part One: Where have we reached?

- The growth rate has been 3.1, 2.5 and 2.7 per cent per annum during 1951-52 to 1964-65 (pre green revolution); 1969-70 to 1987-88 (post green revolution) and 1951-52 - 1987-88 (almost whole period after independence in 1947). The growth in pre-sixties was contributed mainly by bringing new area under cultivation either by cutting forests or previously fallow land. In post sixties, yield increased entirely because of increase in the productivity.
- While the jump from 74.2 million tonnes in 1966-67 (a drought year) to 94.0 million tonnes in 1967-68 after introduction of new technology was indeed very dramatic given the fact that post war economy (wars of 1962 and 1965) was quite sluggish as far as public investments were concerned. However, we had reached a figure of 82.3 million tonnes way back in 1960-61 when monsoon was good. Also, the production level hovered around 100 million tonnes till 1975. It came down to 97 million ton in a year of severe drought in 1972-73 and to 99.8 million ton in another drought year 1974-75.
- The jump of 20 million tonnes in 1975-76, 1982-83 and 1988-89 to take the production levels to 170 million tonnes has been contributed by good monsoon but also better distribution of inputs, improved irrigation infrastructure, technological upgradation, and resurgence of productive potential in Eastern India.

- Per capita availability of foodgrains and pulses has remained around 441 grams per day. It was as high as 480 gms in 1965. Population growth rate has kept a pace with the food production.
- Per capita income of rural producers (1966-77 to 1978-79) has ranged from ₹ 1,627/- ₹, 1,270/- in Punjab and Haryana to ₹ 395/- in Bihar. It ranged from ₹ 600 to 750 in 6 states and ₹ 490 to 600 in 12 states.
- The growth rate of wheat has been 5.1 per cent per annum between 1969-70 to 1987-88 and has been higher or at same level for several other cash crops and vegetables like potato. The oilseeds have a growth rate around 3.7 per cent, rice 2.4 per cent, pulses around 1 per cent, jowar 0.3 per cent and Bajra negative. Thus the green revolution has escaped the rainfed crops affecting the poorest people so far.
- Input productivity has been declining at an alarming pace. With 1970-71 as base of 100, the index has come down to less than 60 in 1987-88. Fertilizer consumption has increased by about 387 per cent during 1970-71 to 1988-89, gross irrigated area has increased by about 30 per cent, power consumption by 88 per cent, pesticides by about 323 per cent during the same period.
  The food production increased by only 57 per cent.
  The non-sustainable nature of the green revolution doesn't need a more telling testimony.
- Despite increasing production in recent years, government has found it difficult to fulfill its procurement current targets to meet obligations of food stock for national security and public distribution system. The larger irrigated land holders have increased their capacity to stock grains.
- The expenditure on drought relief has continuously increased being more than ₹ 2000 crores during 1987. If the bank loans waived or to be waived may be added to it, it may increase by another couple of thousand crores.
- The rainfed crops like oilseeds and pulses have been pushed over to more and more marginal lands making technological change in them all the more difficult.
- Most of the millet, pulse and rainfed oilseed growers being scattered, vulnerable to seasonal fluctuations and living in the regions with low population density have poor access to formal delivery systems for meeting various minimum needs. The market forces are also quite weak.
- Emigration of poor people from drought and flood prone regions and hill areas to cities adds to the problem of urban poverty. This also robs the disadvantaged regions of many able men and women who could have taken some risk to undertake technological

transformation. This is in addition to the misery that children and other family members of these migrants have to suffer.
- The proportion of women headed or managed households, is quite large in these regions. Given conspicuous neglect of women clients by public agencies, poor infrastructure becomes poorer for these families.

HISTORICAL OVERVIEW OF TECHNOLOGICAL CHANGE

There was a strong research interest among Indian Scientists in forties on finding out best way of conjuctive use of organic and inorganic fertilizers. The agronomic ways of weed control were given importance and overall pest and weed problems in any case were less severe. Dr. N R Dhar had pioneered many imaginative experiments on sustainable agriculture in Allahabad.

We begin the story of technological change from 1962, the year India had the first major war with China and realized its poor defense preparedness. National consensus on several macro policy issues primarily aimed at import substituting self-reliant economy started breaking down,

The year 1964-65 was a major drought. The war with Pakistan in 1965 increased the pressure on economy. But stoppage of US shipments of cheap food aid provided the right stimuli for searching alternatives for self-reliance in food sector. It may be recalled that America had by then given considerable aid both for setting up so called community development blocks (of about 80-100 villagers each) all over the country and also for meeting the needs of public distribution and emergency aid for victims of natural disasters. American interest in keeping communists at way was served well.

Distribuion of imported wheat under public employment programmes had created taste and market for wheat even in the non wheat growing regions.* Indian Scientists were aware of the high yield responsiveness of mexican varieties of wheat and IR varieties of paddy. Decision to bulk import the wheat seeds heralded the massive effort for technological change primarily in the Indo-gangetic plain with good rich soil and abundant waters.

Even though the extension machinery was quite weak and so were the markets, the alleged resistance of Indian farmer to a visibly 'viable' technology evaporated in the thin air. The drought of 1965-66 had also witnessed deaths on large scale (the last time it happened) in Bihar. New technology led to high yields and also high incomes but initially to only those who had by 1967, four major events took place:

- The devaluation of rupee and its failure to boost Indian exports - large scale violence in rural areas by left radical group scaled as Nexalite movement.
- Several states had opposition SVD (joint front) governments ruled mainly be the rightist parties - Central government though ruled by congress was critically dependent upon communist party or support.

- By this time, several communist leaders had infiltrated congress party hoping to capture it from within.
- Around the same time, experiment in social control of banking had started. But with split in the ruling party on the eve of election for the president of the country, a new polarized party with relative dominance of socialists ideology emerged on the scene.

A confidential report of the Home Ministry by this time had brought to fore the need for land reforms and pressure for direct attack on poverty if rural violence had to be contained.

Bankers by then had begun to see the potential of mobilizing rural savings accruing on account of technological charge. Government wanted to prove its socialistic ideals and thus nationalised the banks among other measures.

Rural credit also started being pumped into the agricultural sector to improve incentives for technological transformation.

By early seventies, India had to face another war (1971) which transformed the fortunes of the ruling party. Even though politically it was less dependent upon the leftist allies (because it had huge majority in the parliament), the efforts for containing hopes of small and marginal farmers and landless labourer had to be initiated. Several subsidy linked programmes for provision of credit for irrigation, land development and other assets started.

There was a brief spell when several programmes aimed at solving eco-specific problems of the country were launched e.g. Drought Prone Area Programme, Hill or Tribal area development programme etc.

By 1973, the failure of poverty alleviating programmes was obvious. It was also clear that trickle down of growth made possible though new technology may not take place at sufficient pace and scale. The land reforms had got stuck through complicated but easy loopholes left in the acts leading to large scale litigation. The country side was getting restive. The erstwhile socialist colleagues of the them prime minister were also getting disillusioned.

The oil price hike in 1973 and resultant inflation made worse by the drought of 1972-74 resulted in a widespread social unrest. The difference this time was that unrest was not restricted merely to rural areas this time. It spread to urban areas too. By 1974, the country was facing a massive industrial unrest, widespread social discontent and glamour for change. But change that followed was a shock therapy type. Emergency was declared in June 1975 and all fundamental rights were suspended.

Among other things, two or three things had crystalised by now:

- World Bank had extended massive line of credit for mechanisation and minor irrigation;
- technological change was indeed diffusing rather widely leading to perceptible change in food supply;
- Political tilt of ruling party towards the right was being manifested openly.

While Price Commission had been set up in mid sixties, the terms of trade were moving against agriculture. Even though there were a few announcements ostensibly to help poor but it was clear that government had lost contact with people. Emergency excesses led to the change in government. The new political alignment was certainly to the right but also had some committed socialists. The direct attack on the poverty through IRDP (Integrated Rural Development Programme) started in 1978. Agricultural prices were made more favourable and input subsidies continued. Good monsoon coupled with continued supply of new varieties of wheat and paddy helped in achieving new heights in food production. Emphasis remained on input responsive technologies for increasing the yield. By 1983-84, thanks to the recognition of seed distribution as a major bottleneck, attention was shifted to Eastern India where growth rate had been low. Good monsoon coupled with good management of input distribution, food grain production increased again.

By now however, two major signals started appearing:

- The input output ratio was becoming seriously adverse in the agricultural sector.
- Efforts of the government to reduce subsidies to contain budget deficit were not meeting much success.

This government ruled by a coalition of opposite political interests broke up and the old party came back to power. The political change this time had some very clear implications for technology transfer. Despite the drought of 1979, the scientists and the public officials were happy that food production had not dipped too low. The mechanization, which was slowed down due to large scale studies on its labour displacing effects was reintroduced in early eighties. Mounting of inventories in tractor industry was the major reason. But pressure from affluent farmer had also made some difference. The credit constraint had been removed. Fertilizer subsidies continued.

Emphasis was shifting towards high tech approach to agricultural development through reliance on Biotechnology and in fact import of seeds made more easily possible than hitherto. It was realized belatedly that standardized approach for agricultural development relying on the most favoured and well endowed regions may not work any more. The attention started increasing towards regionalization of technological change policies.

But this time next yield barrier was crossed in 1988-89, the non-sustainability of agricultural technology was becoming more and obvious. At the same time need for better soil and water conservation through watershed management in dry regions and restoration of ecological balance in hill areas denuded severely in the past was being felt. But the patience for technologies which could generate surplus in longer period of time but in a sustainable manner was just nt there. There was an appreciation for decentralised planning nor was there any attention towards improving the capacity of agricultural scientists to continue to deliver results. The research funds per scientists in the country had gone down in real terms by as much half and in many state more than that. Budget of Indian Council of Agricultural Research was almost equal to or slightly more than the budget of Department of Biotechnology alone.

Sustainability of the institutions in which research for sustainable agriculture could be done was itself under doubt. The private and public corporate sector (particularly agri-input industries) was thriving on the work of agricultural scientists but was hardly making any investment in improving research productivity.

Where Do we go from Here

The following research needs based on the lessons of Green Revolution in India are unlikely to attract attention within India without considerable pressure from relevant but less powerful constituencies. Some constituencies are historically quite weak such as the ones belonging to graziers or livestock herdsmen, readers of bullocks following extensive systems of resource management; makers of hand tools or bullock drawn implements', small scale manufacturers of herbal non-toxic pesticides; developers of non-monetary input requiring technologies etc.

But then sustainable development requires close involvement of these very constituencies.

- The time of individual oriented technologies is over. More and more emphasis will need to be placed on the group based technologies. The areas where group action is called for include:
  - Synchronization of sowing schedules of certain crop to avoid synchronization with pest reproduction cycle in certain crops where this is a major problem.
  - Watershed management of both arable and non-arable land belonging to individuals as well as villages, forest and revenue department.
  - Synchronized sprays compeld with pest reducing crop rotations and mixtures to minimize use of chemicals in the short run and total elimination in the long run.
  - Biological pest control coupled with water management and drainage.
- Most of the post graduate and even other research particularly in the discipline of Agronomy is concentrated on Inorganic Fertilizer. Very little work is being done on green manuring, conjuctive use of inorganic or organic manure and replenishment of micronutrients through ash and other organic supplements.
- Genetic uniformity in high growth regions has created another major reason for concern. Very few varieties of wheat occupy large areas.

And a large number of rice varieties with IR parents have common genetic source of disease resistance. Diversity in genes cannot be achieved in the short run without making a trade off in yield. Just like Europe and America provide incentives to growers to fallow the land or for cutting the dairy animals, incentives for genetic diversity, and green manuring will have to be provided in India in high growth regions. These will be necessary till alternative genetic sources of resistance and yields are available.

- The shift from crop to trees (horticultural and/or timber) is taking place both for reducing the need for outside labour and also for reaping larger commercial gain by larger farmers. As long as the shift is marginal, nothing much may matter. But once it becomes large scale, the imbalances in food supply may emerge. Agro-forestry for high and slow growth regions thus is an urgent priority.
- The till towards fashionable technologies rather than relevant technologies has to be reversed. Even though we do not have even ten calves produced under field conditions from Embryo-transfer technology, the scarce resources for dairy research and technology transfer are being misallocated towards such technology. Even the conventional technologies like frozen semen (of good local breeds) for upgrading cattle haven't yet been provided at large scale. Shifting gears every now and then on technologies popular in west in costing us a lot. Of course the problem is common to most disciplines and not merely to dairy. If elites in third world recognize merit of third world scholar only/mainly when they get approval/ notice from the west, then is not it natural that the scholars (undoubtedly highly competent) should work on problems fashionable in west.
- Synchronization of maturity was a necessary trait in cereals for improving mechanizations in west. Research on their subject became popular in seventies in India also because many scientists returning from US or UK after training carried on this research. Perhaps alternative ways of combining response to inputs could have been found if searched.
- It should be mandatory for every agri-input industry to contribute a part of its profits to a National Fund for Sustainable Research (NFSR) to be managed by a group of inter-disciplinary scientists, NGOs and committed public servants. NFSR should be used to find long range research programme for identifying low/no external input sustainable technologies.
- One of the causality of the green revolution in India has been the discontinuance or weakening of longitudinal research. Without such research, comprehensive understanding of changes in ecosystem

cannot be achieved. There are hardly any experiments in the country running for 30-40 years to provide basis for developing long range forecasts for different type of technologies (particularly for soil and water conservation, but also for input use).

Unfortunately most foreign aid agencies reinforce such a face long term challenges can't be developed unless basic research on system basis is supported in different agro-climatic regions of the country.

- The continued neglect of dry regions is really criminal. It is a pity that when we think of hill areas or drought prone regions, we conjure image of domestic servant or a cheap labour. National Commission on Development of Backward Areas (1981, Planning Commission, New Delhi) went so far as to say that we should not try to create conditions by which supply of cheap labour for large irrigation projects is affected adversely.
- It is true that far fewer Members of Parliament are elected from dry regions due to low population density and hence much political pressure is not expected in the short run. But it should be acknowledged that low investment in pasture development in dry regions affects the sustainability in other regions directly. For instance,
  - population growth rate of sheep and goat has been several hundred percent compared to 30-60 per cent increase in cattle in most districts in Western India. Poor have to increase weight of low capital requiring enterprise in their portfolio if they have to survive without any support for other possible technologies or enterprises.
  - The cultivation of marginal regions increases siltation rate of seasonal rivulets in these regions further affecting the run off, soil and water conservation. Even snap floods have been experienced in some parts of Rajasthan - a phenomenon unheard of in history.
  - Continued emphasis on crop varieties with high harvest index (grain/straw ratio) has intensified pressure on dry matter supply. The pressure on trees for leaves and energy (twigs) is inevitable. Tractorization in dry regions does damage in two ways, (i) it makes escaping spontaneously sprouted tree seedlings while ploughing virtually impossible (this was possible with bullock ploughing) and (ii) it increases the tendency to cultivate river banks, marginal lands which were earlier left as fallow. Once the soil is loosened and grass roots are removed, the rate of soil erosion increases. Strong winds do bring this sand (and some silt) on to the neighbouring regions. This trend will only increase in future.

  - Mining of ground water in high growth regions has assumed alarming proportion.
- Development of Common Property Institutions for generation of collective restraint is most essential. It is obvious that as the water table/goes down lesser and lesser people are able to mine it. Having lost the battle of land reforms the struggle for water reforms must not be lost.
- The dilemma of continuing malnutrition and poverty among 30-40 per cent of the people having no purchasing power is manifesting in certain paradoxical ways. For instance, wheat was sought to be exported a few years ago when the food stocks reached a level of 30 million ton. It was realised that (a) more and more countries were trying to seek self reliance reducing the demand for wheat at commercial prices, (b) the quality of wheat was not good such that even Russia - a close ally of India refused to buy a few lakhs ton of wheat and (c) the African countries which had need for this grains did not have the ability to pay in hard currency. Instead of using those stocks for launching massive employment programme for reclaiming denuded forest lands and other eroded lands, thinking of export was obviously a short sighted answer.
- The demand of manufactured goods which are income inelastic is unlikely to increase unless the distribution of purchasing power becomes more even.

Green revolution has certainly proved that Indian farmers were no less enterprising than the farmers anywhere else in the world. However, it is also true that the rate and pattern of growth is not sustainable either ecologically or even politically. Some of the tensions in high growth regions are arising because of the inability of state to continue the subsidy at the rate at which these were available earlier. On the other hand that deterioration in the level of living of the people in drought prone regions and hill areas is leading to demands for separate states and more autonomy. There is no denying the fact that next round of violence could be in the backward regions if neglect continues to be what it has been. The recent trend towards inviting multinational corporations for initiating agribusiness model of contract farming in the high growth regions cannot but bring the experience of Latin America and Phillipines into India. It is a pity that planners cannot see that.

Index

A

Accumulation 46, 48, 64, 70, 100, 105, 154, 205
Adaptation 9, 98, 99, 105, 178, 179, 180, 186, 204, 206, 207, 208, 216, 217, 250, 252, 266, 267, 268, 271, 273, 275, 276, 277
Agronomical 103, 104, 276
Agropastoral 163, 164, 165, 166, 168, 175
Arrangements 25, 26, 140, 160, 164, 165, 191, 251

B

Barbuda 106, 107, 108, 110, 111, 112, 113, 114, 115, 116, 117, 118, 120, 121, 122, 123, 124, 125, 126, 129, 130, 131, 132, 133, 134, 136, 137, 139, 140, 141, 142, 144, 146, 147, 148, 149, 150, 153

C

Catchment 62, 89, 97, 129, 138, 166, 184, 185, 188, 190, 194, 221, 222
Communities 9, 10, 20, 21, 36, 38, 40, 41, 51, 53, 55, 56, 59, 62, 64, 65, 68, 69, 70, 71, 123, 154, 155, 156, 159, 161, 162, 164, 165, 177, 178, 179, 181, 184, 191, 194, 245, 252, 256, 261, 277, 278, 279
Conflict 4, 20, 21, 22, 42, 46, 47, 55, 56, 61, 62, 63, 64, 65, 157
Consecutive 130, 188, 226
Coordination 23, 24, 25, 95, 99, 158, 178

D

Demographic 5, 251, 252
Discussions 41, 109, 138, 251, 252

E

Evaporation 85, 86, 88, 94, 107, 111, 112, 113, 139, 152, 187, 196, 210, 211
Experiments 33, 85, 86, 87, 88, 209, 217, 283, 288

G

Geology 113
Globalisation 49, 51
Greenhouse 76, 77, 86, 99, 270

H

Hydrological 2, 3, 24, 73, 80, 83, 85, 86, 87, 88, 89, 90, 91, 93, 94, 108, 109,

112, 113, 117, 122, 125, 128, 130, 132, 144, 148, 149, 150, 151, 152, 154, 156, 157, 190, 194, 195, 196, 212, 221, 222, 224, 249, 250
Hydropower 45, 271

I

Implementation 9, 21, 23, 24, 26, 27, 43, 52, 137, 157, 162, 190, 191, 277
Influence 41, 47, 65, 78, 86, 87, 100, 103, 110, 111, 112, 113, 134, 140, 141, 142, 145, 152, 168, 171, 188, 191, 226, 229, 239, 244
Institutional 19, 21, 23, 24, 25, 26, 27, 28, 52, 53, 54, 137, 138, 140, 155, 156, 157, 158, 162, 167, 191, 192, 268, 273, 277, 278, 279

L

Livelihoods 9, 38, 46, 47, 48, 49, 51, 52, 53, 54, 55, 56, 57, 58, 60, 61, 62, 63, 64, 65, 66, 67, 69, 154, 155, 156, 160, 161, 268, 269, 271, 272, 278, 279, 281

M

Meteorological 2, 3, 10, 13, 24, 27, 39, 54, 82, 83, 84, 88, 89, 91, 92, 99, 101, 119, 120, 121, 122, 140, 156, 163, 167, 168, 171, 172, 175, 176, 247, 249, 250, 256, 273
Methodology 16, 23, 106, 196
Mitigation 4, 5, 6, 8, 12, 19, 20, 24, 26, 27, 28, 30, 31, 32, 35, 36, 40, 41, 42, 43, 69, 73, 77, 106, 107, 109, 130, 138, 140, 148, 149, 150, 154, 155, 156, 157, 159, 162, 268, 271, 273, 275
Monitoring 4, 6
Morphology 208

P

Pastoral 35, 36, 37, 38, 39, 40, 41, 42, 62, 63, 64, 65, 154, 155, 156, 157, 159, 160, 161, 162, 163, 164, 165, 166, 168, 172, 173, 175
Photosynthesis 101, 102, 103, 104, 204, 205, 217
Predictands 75
Preparedness 2, 4, 5, 6, 8, 11, 18, 19, 20, 23, 26, 40, 43, 108, 138, 140, 155, 156, 158, 219, 221, 283
Preventability 72

R

Recommendations 23, 80, 95
Regional 14, 20, 21, 22, 23, 24, 25, 26, 27, 47, 52, 53, 72, 77, 78, 79, 82, 83, 84, 85, 86, 87, 88, 89, 90, 91, 92, 97, 98, 173, 174, 176, 181, 196, 197, 221, 268, 285

S

Socioeconomic 2, 3, 223, 247, 262
Strengthening 7, 20, 21, 24, 26, 36, 65, 166, 176, 177, 178
Supplementary 35, 36, 38, 39, 42, 159, 160, 161, 164, 169, 218

U

Urbanisation 47, 49, 67

V

Vulnerability 1, 3, 5, 9, 23, 25, 26, 46, 47, 55, 56, 58, 59, 60, 61, 63, 64, 68, 69, 70, 106, 107, 108, 116, 119, 120, 123, 124, 125, 128, 136, 137, 138, 139, 140, 141, 142, 143, 144, 145, 146, 147, 148, 150, 163, 182, 224, 246, 247, 248, 250, 252, 259, 260, 261, 266

W

Watershed 7, 78, 117, 118, 119, 120, 125, 126, 131, 132, 144, 145, 146, 147, 148, 149, 150, 151, 189, 190, 191, 193, 199, 202, 285, 286